Computational Design and Robotic Fabrication

Series Editor

Philip F. Yuan, *Tongji University, Shanghai, China*

This open access book series includes compilations of selected papers from the International Conference on Computational Design and Robotic Fabrication. The books focus on novel techniques for computational design and robotic fabrication. It not only aims at the most recent research results from the key scholars in the computational design and robotic fabrication, but also offers an in-depth examination of intelligence in design and construction industry, referring to the invention and application of machine intelligence in architecture.The proceedings in this series are related with Sustainable Development Goals 11 (Sustainable Cities & Communities) and 9 (Industry Innovation & Infrastructure).

The contents make valuable contributions to academic researchers, designers, and engineers in the industry. As well, readers will encounter new ideas about understanding intelligence in architecture.

Indexed by EI Compendex and Scopus.

Jiayan Fu · Xuan Zong · Philip F. Yuan
Editors

AI for Architecture

Proceedings of the 5th International Conference on Computational Design (CCD 2024)

Springer

Editors
Jiayan Fu
College of Architecture and Urban Planning
Tongji University
Shanghai, China

Xuan Zong
College of Architecture and Urban Planning
Tongji University
Shanghai, China

Philip F. Yuan
College of Architecture and Urban Planning
Tongji University
Shanghai, China

ISSN 2731-9040 ISSN 2731-9059 (electronic)
Computational Design and Robotic Fabrication
ISBN 978-981-95-0973-7 ISBN 978-981-95-0974-4 (eBook)
https://doi.org/10.1007/978-981-95-0974-4

This work was supported by National Natural Science Foundation of China.

This Springer imprint is published by the registered company Springer Nature Singapore Pte Ltd.
The registered company address is: 152 Beach Road, #21-01/04 Gateway East, Singapore 189721, Singapore

Preface

Currently, global digital transformation is accelerating industrial change at an unprecedented pace. Amidst the Fourth Industrial Revolution, artificial intelligence has emerged as a key technological driver, reshaping value creation systems and paradigms for industrial upgrades within the construction sector. Innovative approaches, such as computational design and robotic construction, are facilitating the ontological development of architecture, enabling continuous iterations that promote breakthroughs in multimodal capabilities—from environmental perception to intelligent design and construction—within the realm of architectural design and theory. The synergistic integration of deep neural networks and swarm intelligence algorithms has significantly improved the real-time intelligent sensing, precise control, and collaborative operation capabilities of construction systems, offering revolutionary solutions for optimizing efficiency and enhancing quality in engineering projects. This technological evolution points toward a comprehensive intelligent reconstruction of construction processes.

In China, the construction industry is grappling with dual pressures of transformation: on the one hand, demographic shifts have rendered the traditional labor supply model unsustainable; on the other hand, the strategic goals of achieving "carbon peak and carbon neutrality" impose heightened demands for greener construction processes. In this context, the advancement of intelligent design and construction technologies that seamlessly integrate intelligent sensing, digital twins, and low-carbon processes has become essential for overcoming industry bottlenecks.

Since its establishment, the Computational Design Committee of the Architectural Society of China has been committed to promoting digital and intelligent innovation in the construction field. The 2024 Computational Design Professional Conference, themed "Computational Design: AI for Architecture," highlights how computational design has transcended the limitations of manual drawing and empirical judgment in traditional architectural design processes. The rapid rise of artificial intelligence has infused computational design with vitality and intelligent elements.

This conference is guided by the Architectural Society of China and co-hosted by the Computational Design Committee of the Architectural Society of China and the College of Architecture and Urban Planning at Tongji University. It is co-organized by the School of Architecture and Design at Harbin Institute of Technology, DigitalFUTURES, Shanghai Building Decoration Engineering Group Co., Ltd., D5 Render, Tongji Architectural Design (Group) Co., Ltd., Shanghai Tongji Green Building and Prefabricated Construction Engineering Technology Co., Ltd., Shanghai Digital Architecture Fabrication Technology Center, and the Digital Architecture Branch of the Architectural Society of Shanghai China. The conference aims to promote the sustainable development of computational design theories, methodologies, technical tools, and engineering practices. The event is divided into four parts: keynote presentations, corporate presentations, parallel sessions, and workshops, attracting over 500 scholars from related fields both domestically and internationally.

The highlight of combining academic conferences with workshops lies not only in the expansion of theoretical boundaries but also in the breakthrough of industry-academia-research collaborative innovation models. The deep integration of academia and industry has given rise to cross-scale technology transformation pathways. Cutting-edge algorithms from university laboratories are rapidly iterated into engineering solutions through corporate engineering research centers, while practical data from construction sites drives paradigm updates in fundamental research. This two-way empowerment mechanism not only achieves the organic integration of traditional construction techniques and autonomous intelligent technologies but also validates the closed-loop logic of parametric design transitioning to industrial implementation, marking computational design's leap across the "laboratory-construction site" divide.

Among the highlights of this conference, a total of 281 papers were submitted, with 51 accepted for presentation. The event featured 11 parallel sessions, organized into five thematic tracks: Computational Design and Construction, Computational Urban Science, Computational Urban Design, Computational Landscape, and Computational Media and Art. Scholars from 16 prestigious institutions, including Tsinghua University, Tongji University, Southeast University, Tianjin University, Chongqing University, Harbin Institute of Technology, South China University of Technology, Xi'an University of Architecture and Technology, China University of Mining and Technology, Zhejiang University, Nanjing University, Hunan University, Shenzhen University, Hefei University of Technology, Inner Mongolia University of Technology, and Nanjing Tech University, shared their research findings. These sessions provided a robust platform for inspiring researchers and opening new avenues of inquiry, playing a positive role in advancing academic research and innovation in related fields.

This proceedings volume compiles 17 outstanding papers. It is hoped that these contributions will inspire more interdisciplinary collaborations, driving the construction industry toward a deeper integration of "digital intelligence" and "sustainability" under the strategic goals of "carbon peak and carbon neutrality." Guided by architectural design thinking, these efforts aim to provide Chinese insights for the high-quality development of global, virtual, and extraterrestrial human habitats.

Philip F. Yuan

Shanghai, China
March 2025

Committees

Hao Shimeng, Beijing University of Civil Engineering and Architecture
Shen Yuchi, Southeast University
Cao Ting, Harbin Institute of Technology (Shenzhen)
Guo Zhe, Hefei University of Technology
Hu Biao, Hunan University
Zhang Rongpeng, Hunan University
Deng Qiaoming, South China University of Technology
Lin Zhenghao, South China University of Technology
Tong Ziyu, Nanjing University
Shi Xinyu, Qingdao University of Technology
Huang Weixin, Tsinghua University
Lv Shuai, Tsinghua University
Shi Shaohang, Tsinghua University
Fan Yating, Shenzhen University
Wang Haofeng, Shenzhen University
Wan Da, Tianjin Chengjian University
Fu Jiayan, Tongji University
Li Wenjing, Tongji University
Liu Chao, Tongji University
Xu Xiaoqing, Tongji University
Ye Yu, Tongji University
Wang Likai, Xi'an Jiaotong-Liverpool University
Zheng Hao, City University of Hong Kong
Yu Jiahao, University of Hong Kong
Wang Shuaizhong, Chinese University of Hong Kong
Xu Weishun, Zhejiang University

Contents

Computational Urban Science

Computational Urban Design

Computational Landscape

Computational Media and Art

Computational Design and Intelligent Construction

Generative AI in Architectural Design

Xinhui Xu[1], Zhengcheng He[1], Tinghao Feng[2], and Philip F. Yuan[1,3](✉)

[1] College of Architecture and Urban Planning, Tongji University, Shanghai, China
philipyuan007@tongji.edu.cn
[2] KTH Royal Institute of Technology, Stockholm, Sweden
philipyuan007@tongji.edu.cn
[3] Tongji Architectural Design (Group) Co., Ltd., Shanghai, China

Abstract. Generative Artificial Intelligence has brought a huge impact on architectural design. Architectural analysis and generative concepts together present a new opportunity to apply generative AI techniques at the early design period. This work reviews research on generative AI applied in architectural design to enhance the understanding of architectural advancement. Several challenges arise while doing such generation such as, building a knowledge graph, "Black-Box" Issue, computation limitation. In this paper, we present an intuitive review about such possibilities for generative AI in architectural design. Indeed, we provide innovative directions for generative design in the future.

Keywords: Generative artificial intelligence · Generative design · Generative architectural design

1 Introduction

The rapid advancement of Generative Artificial Intelligence (GAI) has placed it at the forefront of technological innovation, particularly in architectural algorithms and generative design. This technology has the potential to redefine architectural design, introducing a new paradigm driven by AI. Leveraging vast open-source architectural datasets and unprecedented computational power, GAI is beginning to challenge manual design and parametric modeling, signaling a shift toward an "AI-driven era" in architectural design. While generative AI has achieved remarkable success in various fields—including artistic creation, music composition, virtual assistants, and narrative generation—its application in architectural design remains in its early stages. Consequently, exploring multi-scenario applications has become a key strategy for driving architectural paradigm shifts in the AI-driven era.

2 Early Generative Concepts and Methodologies in Architectural Design

Mario Carpo categorizes the progression of digital architecture into four distinct phases [1]: an initial phase (False Start) that began in the 1950s; a foundational period (Real Beginnings) that encompassed the 1990s; the second digital turn that occurred around the

J. Fu et al. (eds.), *AI for Architecture*, Computational Design and Robotic Fabrication,
https://doi.org/10.1007/978-981-95-0974-4_1

2000s; and the ongoing AI-driven era (AI Reborn) that emerged around the 2010s. Early generative design research focused on search-based methods such as cellular automata, genetic algorithms, swarm intelligence, and fractal algorithms. These methods usually establish a clear generative framework at the beginning—for example, defining the four corners of a building in a plane and optimizing their spatial configuration based on sun angles.

However, architects often do not have such firm ideas or a clear-cut rationale in the early stages of a project. Architectural design is a complex and lengthy process. The design process usually starts with collecting relevant cases and gradually forms a vague concept that supports the architectural plan. In addition to internal factors such as aesthetics, many external influences, regulations, and other forces are equally important in the design process. These multi-dimensional factors make it impossible to complete the design in one iteration. The final architectural form cannot be established in a clear design logic and needs to be gradually improved through multiple iterations and scheme screening. Early generative design methods prematurely restricted architectural design to a framework, resulting in a lack of creativity in the generated results.

The Generative Adversarial Network (GAN) that emerged in 2014 is an important milestone in GAI, further broadening the scope of creativity in generative architectural design. The development of GAI can be summarized into three major stages. Early (1950s to 1990s): technical limitations, limited to small-scale experiments; mid-term (1990–2010): deep learning emerged in computing resources, giving rise to prototypes of GAI technology, such as the Deep Belief Networks (DBNs) that appeared in 2006; Development stage (2010 to present): GAN has completely changed GAI technology, and a large number of dynamic content generation methods based on artificial intelligence have emerged.

At present, GAI has been applied to various fields of architecture, such as data quality enhancement, building layout generation, 3D printing path generation, etc. With the development of diffusion models and GPT series, GAI has released huge potential, and over time it will drive a paradigm shift and fundamentally change the construction industry. Due to the dynamic and complex nature of architectural projects, design solutions are closely related to constant adjustments. In addition, due to the involvement of multiple interrelated feedback systems, changes in specific conditions often lead to unpredictable changes, which pose a challenge to the application of traditional machine learning methods. A better understanding of GAI will help improve its effectiveness, quality, and scope of application in architectural design.

3 Generative AI in Architectural Design

3.1 Enhancing Architectural Design with Generative AI

AIGC is one of the most advanced fields in contemporary AI technology, which is able to generate new data in various fields such as text, images, audio, etc. This ability refers to the automatic generation of content by AI, which can imitate human creativity to create novels, artworks, paintings and even music. AIGC mainly relies on generative AI technology, in which generative modeling aims to infer the underlying data distribution. For example, the joint probability distribution of input and output $P(X|Y)$ or $P(X)$, where X is usually from a high-dimensional space. By doing so, the generative model generates new synthetic samples under given conditions. For instance, when provided with a condition Y, the model generates new featreure-observation pairs (X, Y) or new observation values X, thereby automating the creation of new designs.

In architectural design, AIGC can analyze vast amounts of architectural data and design patterns to generate multiple design proposals, optimizing them according to specific requirements such as functional layout, structural stability, energy efficiency, and aesthetics. The application of AIGC technology not only reduces the burden of repetitive tasks for designers but also drives the architectural industry toward smarter and more sustainable development.

3.2 Mainstream Technologies

In architectural design, image and model generation technologies are at the forefront. These technologies are primarily structured into three algorithmic frameworks [2]:

- **Diffusion Models**: These models are widely used for image generation. They capture image data by modeling the noise addition process in latent space. Specifically, they employ a Markov chain trained through variational inference, followed by reverse diffusion to generate realistic images [3]. Diffusion models power leading image generation algorithms and applications such as Stable Diffusion, DALL-E, and Midjourney.
- **Generative Adversarial Networks**: GANs consist of two neural networks: a generator and a discriminator. The first network, the generator, generates fake samples, while the second network, the discriminator, evaluates whether the input samples are real. Through an adversarial process, the generator continuously improves its ability to generate realistic samples, while the discriminator continuously improves its ability to distinguish between real and fake samples [4]. This iterative process leads to the generation of increasingly realistic samples that closely resemble the real data distribution. GANs are not only simple in structure, but also able to generate completely new samples, which has made them the dominant method for a long time. They include some of the more well-known variants, such as conditional GAN (cGAN), pix2pix, StyleGAN, and CycleGAN.
- **Variational Autoencoder (VAE)**: VAEs learn low-dimensional representations of input data by encoding them into a compressed latent variable space and then reconstructing the original data from this compressed representation. By using probabilistic methods in the encoding and decoding process, VAEs can capture the underlying structure and variations in the data [5]. They can generate new data samples from the learned latent space and are commonly used in tasks such as anomaly detection, data compression, and image and text generation.

4 Current Applications of Generative AI in Enhancing Architectural Design

4.1 Site Analysis

- Building Recognition

In order to identify buildings in the real world, many tasks in architectural design, such as energy consumption simulation of existing buildings, building renovation, and urban renewal, require accurate extraction of facade features. Traditionally, these tasks often rely on physical measurements, a process that is time-consuming and laborious. The image conversion and segmentation capabilities of generative AI can quickly extract element features from architectural images, thereby simplifying calculations. A series of studies have developed frameworks based on semantically driven edge detection or convolutional neural network (CNN)-based learning-driven methods that can automatically detect the window-to-wall ratio of facades [6], identify building heights, and predict building areas in street view images. Another method improves existing instance segmentation algorithms, such as SOLOv2 and Mask R-CNN, to perform facade window segmentation. This method eliminates image distortions and occlusions in building facades, enabling the extraction of facade features. This foundational data is essential for applications such as building energy consumption simulations and urban renewal planning [7].

In architectural drawing recognition, technical drawings provide comprehensive building information, but manual calculations based on these drawings have traditionally been required. Generative AI-powered drawing recognition technologies offer significant advantages for interior reconstruction and high-precision restoration. Current approaches utilize YOLO-series models and R-CNNs for object detection, while CNNs are employed to classify engineering drawings, such as electrical or mechanical schematics. Additionally, GAN-based methods focus on recognizing structural, decorative, and functional elements in floor plans [8], enabling automated generation of Building Information Models (BIM).

- Context Perception

Generative AI can be integrated with architectural space analysis to extract spatial features and predict various evaluation metrics based on historical and real-time data. Traditional simulation tools are often complex, computationally expensive, and time-consuming. However, generative algorithm models enable significantly faster computations, facilitating rapid evaluation of form, space, and layout performance in early-stage design optimization. Furthermore, trained models can predict design outcomes with greater accuracy, improving decision-making in architectural projects.

For behavior performance analysis, generative AI models, such as GANs (e.g., pix2pixHD), can generate activity heatmaps based on architectural floor plans. These heatmaps provide insights into the distribution of activities and behavioral patterns within built environments. For example, a research team led by Hao Zheng trained a pix2pixHD-based model to generate activity and vitality heatmaps for different sections of the built environment. This method helps explore the relationship between architectural design and cyclist behavior, allowing for targeted improvements in regional vitality [9]. Similarly, the ARD team at Foster and Partners utilized parametric modeling to generate thousands of floor plan datasets. They then applied spatial and visual connectivity analysis using a fully convolutional U-Net network and Pix2Pix, enabling detailed assessments of spatial and visual connectivity [10].

In environmental performance computation, generative AI is commonly used as a surrogate modeling tool. For instance, daylight simulation surrogate models can be trained by pairing floor plans with geometric and daylighting data. These trained models predict daylight performance in architectural designs, enabling rapid evaluation during the early design stages [11]. Additionally, some studies focus on multi-modal feature fusion, combining pixel-based and vector-based representations of plans, elevations, reflectance, and transmittance. This fusion mechanism enhances the speed and accuracy of daylight performance assessments [12]. Furthermore, research integrating GANs with existing building performance models and context-aware design data allows for the generation of optimized building performance models guided by specific performance goals [13].

4.2 Design Generation

- Concept Generation

As image generation technologies advance, more architects and design firms are integrating generative AI into the early design process to create concept sketches, accelerating prototype iterations. Key algorithms in this workflow include StyleGAN, CycleGAN, Stable Diffusion, and ControlNet. The process involves gathering reference images, compiling datasets, and training or fine-tuning models. Once trained, these models generate architectural concepts through various methods, such as text-to-image generation, image-to-image generation, and inpainting. Several architectural firms have explored these approaches, including Coop Himmelb(l)au's *The Cloud Augmentation* project [14], and Zaha Hadid Architects' *Zaha "most" Image Set* [15].

Text-to-Image Generation: Text-to-image generation is one of the primary cross-modal generative models, particularly useful in the early design phases when design intentions are still evolving. This method allows architects to quickly generate concept images based on textual prompts. For example, Manas Bhatia's conceptual project *AI X Future Cities* utilized MidJourney to generate architectural designs using prompts such as *Future Tower*, *Utopian Technology*, *Symbiosis*, and *Bioluminescent Materials* [16]. Architect Karisma Shoker employed MidJourney to visualize an entrance into the mythical city of Atlantis [17]. Green Clay Architecture used AI tools to merge modern engineering with Persian traditional aesthetics, designing a stadium featuring an integrated structure of concrete, wood, and glass [18].

Despite its potential, text-to-image generation has limitations in precise text controllability, making it most suitable for projects where architects have not yet developed a fully defined concept.

Image-to-Image Generation: Image-to-image generation enhances early-stage design control by using one or multiple images as content and style guides to produce refined visualizations. This approach is commonly applied in two key scenarios:

A dataset of paired sketches and architectural images is first established, followed by training CycleGAN and cGAN models to automatically transform sketches into detailed architectural renderings.

Given early-stage 3D models created in software like SketchUp or Rhino, ControlNet can generate high-quality architectural renderings. The input data can vary widely, from hand-drawn sketches and model screenshots to real-world architectural images or even non-architectural object compositions. The type of input and the selected algorithm significantly influence the final rendered visualization. Additionally, inpainting can modify specific image areas, allowing generative AI to re-render selected sections based on user-defined inputs.

Spatial Layout Planning: Beyond conceptual generation, generative AI contributes to site planning and internal building organization by facilitating spatial layout design. Architectural spatial layout planning requires an understanding of topological relationships, and generative AI provides a scalable solution by integrating diverse data sources. Early spatial layout automation mainly relied on methods such as shape grammars and genetic algorithms, but these methods take a long time to find the optimal solution. At present, the general method of generating spatial layout is to input the existing site boundaries or conditions, represent the spatial relationship through a graph structure, and then use GAN or diffusion models to train these models to guide the design process. For example, in the study of Building-GAN, a voxel graph architectural representation was proposed, and a cross-modal graph neural network was used as a generator. It was trained in a generative adversarial framework to achieve three-dimensional space partitioning with functional annotations [19]. In the functional layout design of a single building, cGAN divides the empty room into several functional parts and trains a fully connected network to place furniture in these functional parts to complete the furniture layout of the room [20]. In addition, spatial planning generation is also used to search for similar spatial layouts. These methods usually extract the topological structure of floor plan or model data and retrieve matching designs from a database. For example, by inputting an image or sketch, the query type can be automatically determined and similar floor plan images can be retrieved from a database [21].

- Drawings Generation

The generation of floor plans is one of the most widely applied uses of generative AI in architecture. These applications typically employ GANs and Stable Diffusion models to convert architectural floor plans into data representations, which deep learning algorithms then process for training and generation. Standard feature extraction methods include CNNs, where floor plans are treated as pixel-based images and analyzed in blocks, and Graph Neural Networks (GNNs), which represent floor plans as graph structures. More recently, self-attention mechanisms have been explored for feature extraction.

Floor Plan Generation: Early applications of generative AI for floor plan generation relied on GANs, often using style transfer models like StyleGAN and supervised learning approaches such as cGANs. In 2018, Wei-Xin Huang and Hao Zheng developed a system using pix2pixHD to recognize and generate residential floor plans [22]. In 2020, Stanislas Chaillou introduced *ArchiGAN*, a Pix2pix-based GAN model that created a generative model stack for designing residential floor plans. However, CNN-based methods had limited real-world applications due to their reliance on rasterized images, which made modifications and precise feature extraction difficult.

To overcome these limitations, researchers shifted from pixel-based representations to GNNs, enabling better identification of architectural patterns and relationships within floor plan structures. This shift aimed to balance creative freedom with controlled floor plan generation while providing valuable insights for research. In 2021, Nelson Nauata introduced *House-GAN++* , which utilized a graph-constrained GAN and cGAN model to encode functional relationships in house floor plans [24]. Subsequently, Mohammad Amin Shabani leveraged graph structures in diffusion models to develop *HouseDiffusion*, which generates vector-based floor plans [25]. GNN-based approaches have proven effective in encoding the spatial relationships, location, and scale of rooms, marking a significant advancement in AI-driven architectural design.

Façade Generation: Generative AI can assist in façade design by providing architects and clients with real-time previews, allowing greater control over style and massing relationships while significantly reducing design time. Façade generation also supports performance assessments based on façade attributes. Algorithms such as StyleGAN, CycleGAN, and ControlNet are commonly used, processing input sketches or four-view models to generate detailed façade designs.

This technology is particularly useful for façade renovation and urban streetscape generation. For example, GANs can generate a façade design for a building by referencing the architectural style, dimensions, and openings of two adjacent structures [26]. A research team at Harbin Institute of Technology compiled a dataset of façades from Harbin Central Street, segmenting walls, doors, windows, and cornices [27]. Using CycleGAN, they created façade images and segmented images for training, enabling the generation of façade designs from segmented input data [27]. As diffusion models continue to evolve, façade generation technologies are expected to improve further, leading to more sophisticated applications.

Structural Drawing Generation: The generation of architectural structural drawings typically builds upon existing floor or sectional drawings, integrating design data, mechanical principles, and experiential rules to refine structural layouts. Since structural design rules vary significantly across different building types, specific neural networks must be tailored to each dataset.

Research in this area has primarily focused on conventional structures, employing CNNs or GNNs to extract building elements from architectural drawings. These methods are often combined with GANs, attention mechanisms, or Long Short-Term Memory (LSTM) networks to generate layouts for shear walls, frame structures, and other load-bearing components. For instance, a trained GAN can rapidly generate shear wall designs based on an input floor plan for reinforced concrete shear wall structures and steel frame support systems.

- 3D Models Generation

3D models convey fundamental aspects such as building volume, proportions, and scale. Generative AI can streamline the 3D modeling process, reducing the manual effort required in early design stages. Architectural forms are typically represented using two main approaches: (1) Explicit Representations—Point clouds, voxels, meshes, and Industry Foundation Classes (IFC). (2) Implicit Representations—Neural Radiance Fields (NeRF) and 3D Gaussian Splatting. The choice of representation influences the generative AI techniques used, including diffusion models, GANs, and VAEs.

Early Developments in 3D Model Generation: Initial research in AI-driven 3D model generation was limited by algorithmic constraints and computational power. Early approaches focused on generating continuous sectional variations from 2D images, which were then assembled into 3D forms through techniques like voxelization, point cloud interpolation, and geometric stretching. StyleGAN and style transfer methods were used to create smooth transitions between different architectural forms, which were later reconstructed into 3D models.

Text-Based and Image-Based 3D Model Generation: With advancements in generative AI, researchers explored two key approaches to 3D architectural modeling: (1) **Text-Based Generation**: Algorithms such as *Text2Shape, Point-E, and Shap-E* generate 3D geometric forms based on textual prompts. These models convert text descriptions into architectural components, which can then be optimized using voxelization to meet design constraints. (2) **Image-based Generation**: Instead of textual inputs, this approach learns directly from images, such as a Large Reconstruction Model (LRM) for 3D.

Semantic Integration in 3D Architectural Models: To align generative AI with architectural intuition, researchers have explored data formats that better represent building semantics. One approach involves integrating AI with BIM. For example, Wei-Xin Huang et al. developed a 3D modeling action log dataset within the BIM framework and trained a Transformer model to predict 3D modeling actions [29]. Other researchers have used Mask R-CNN to extract architectural components (e.g., walls, doors, windows, columns) from floor plan images, mapping them to IFC formats for reconstruction within BIM [30].

Another innovative approach by Matias del Campo applied 3D graph convolutional networks (3D-GCNs) to architectural modeling. This method integrates generative AI into an optimization framework that deforms and refines input 3D models, improving adaptability and efficiency in architectural design [31].

4.3 Process Exploration

- Design Process Framework

Generative AI has significantly reshaped the architectural design process by fostering a human–machine interaction model tailored to architects' needs. AI-driven design systems are structured to adapt dynamically to evolving architectural contexts. Within this framework, generative AI facilitates various functions, including sketch-to-render generation, 3D models generation, trend analysis, and text-to-image or text-to-report generation.

The design process framework often varies depending on the specific task. Giuseppe Bono's research 'Text-to-Building' articulates a design process framework shaped by generative AI, where Shap-E initiates 3D geometries, evolving through computational design into architectural structures. This framework, formed through Generative model integration, adapts dynamically to specific spatial exploration and structural optimization tasks [32]. Anca-Simona Horvath combined the Evolo competition design process to introduce the use of text-to-text, text-to-image, and image-to-image models in the design concept stage, analyzed the three levels of language: architectural discourse, programming discourse, and annotation, and explained the application of these levels in the design process [33]. In addition, some studies have proposed a fast design workflow: starting from black and white sketches to generate conceptual floor plans, which are then processed into depth maps and converted into three-dimensional point cloud models using modeling software. Multiple perspectives of these models are then used to generate rendering images, thereby achieving rapid conceptualization and controllable generation of architectural renderings based on text descriptions.

- Design Tool Platforms

Design tool platforms using artificial intelligence technology provide architects with efficient and intelligent design solutions. Liu Nianxiong's team developed a smart residential design platform driven by green building energy efficiency, which supports design information input, parametric design generation, energy consumption assessment and optimization, and comprehensive decision-making [34]. Li Biao's team developed SIM-Forms, which integrates parametric modeling, quantitative index calculation, and artificial intelligence image generation. In the early stages of design, designers can generate various architectural forms through simple operations on the web platform to obtain key quantitative indicators and conceptual designs [35]. Wanyu He's team developed the XKool artificial intelligence cloud product, which combines text-to-image and image-to-image models to help architects inspire architectural design in the early stages [36]. Tongji University's AIDC team developed *FUGenerator*, which serves the process from generation to construction and exports images for 3D printing.

Additionally, many architectural firms are investing in AI-powered design platforms. Emerging tools such as *ARCHITEChTURES, Ark Design AI,* and *Maket* are rapidly expanding the capabilities of generative AI in architectural design.

- Design and Fabrication Cohesion

Bridging the gap between generative design and construction remains a key challenge in architecture. Architectural drawings and models often lack the detailed information required for seamless integration into construction workflows. Generative AI can address this issue by extracting essential design features and guiding robotic construction processes.

For instance, a study developed an automated bricklaying framework using cGANs. Given a wall shape as input, the system generates optimal bricklaying patterns, which are then converted into robotic assembly instructions [37]. This marks a significant step toward fully integrated generative design and robotic construction, paving the way for large-scale, customized fabrication.

Despite these advancements, research on generative AI in construction remains limited. While much attention has been given to optimizing generative design algorithms and enhancing design diversity, fewer studies have addressed the practical translation of AI-generated designs into construction processes. In particular, systematic studies and empirical analyses on automated construction using generative AI are still lacking. Future research should focus on refining AI-driven construction methodologies to bridge the gap between architectural design and real-world implementation.

5 Key Issues in Generative AI-Assisted Architectural Design

5.1 Challenges in Building a Professional Knowledge Graph

With the rapid development of artificial intelligence and big data technologies, the construction of domain knowledge bases has become increasingly important in the field of architecture. A knowledge base system that centrally stores, classifies, manages, and retrieves professional knowledge will provide necessary semantic information for various intelligent applications, enabling these applications to retrieve relevant information at any time, conduct question-answering, prediction, design generation, and training faster, thereby improving the accuracy and efficiency of decision-making. However, there are many challenges in building an architectural expert knowledge base, including the complexity of architectural semantics, the difficulty of knowledge acquisition, and the lack of standardized data. The information contained in architectural drawings is often too complex, and accurately identifying, extracting, and learning sparse key features is essential for generating high-quality designs.

Given the comprehensiveness of architectural information, this field covers multiple fields such as architectural design, structural engineering, materials science, and construction management, each of which has its own unique knowledge system and terminology. These knowledge systems include both quantitative data, such as building specifications and material performance parameters, and qualitative information, such as design style and spatial layout. Due to the diversity and interconnectedness of these information types, effectively integrating them into a unified knowledge base has become one of the main challenges in knowledge base construction. In addition, cross-platform data security, privacy protection, and intellectual property issues further complicate the construction of knowledge bases.

Regarding the difficulty of knowledge acquisition, the rapid growth of Internet data requires the extraction of relevant knowledge from a large amount of literature, databases, and expert experience. This process is not only time-consuming and labor-intensive, but also requires researchers to have deep domain expertise. In addition, architectural knowledge usually exists in unstructured text formats, such as design specifications, engineering logs, and meeting minutes. Effectively extracting useful information from these texts and converting them into structured knowledge representations remains a major challenge in knowledge base construction.

5.2 "Black-Box" Issue and Decision Transparency

Architectural design often requires precise numerical descriptions involving complex networks of components. However, the use of generative AI (which relies on deep learning models based on fitting principles) leads to an opaque decision-making process, often referred to as the "black box" nature of these models. This lack of explainability affects the efficiency and quality of architectural design. As AI systems are increasingly used in the construction sector, this "black box problem" has raised widespread concerns about safety, compliance, and accountability.

The construction industry obtains data from multiple sources, including design drawings, sensor data, and market information. However, these data sets often lack effective learning methods, further exacerbating the opacity of AI systems. In addition, architectural planning often requires optimizing multiple objectives simultaneously, but AI systems lack transparent decision logic when generating design solutions. Such solutions must consider various factors, such as physical properties, climate conditions, and other contextual data, making it difficult to trace the internal logic or understand the impact of each input condition on the final output. In addition, the construction industry has high requirements for safety and reliability, and the "black box" nature of generative AI systems may cause important safety standards and local risks to be overlooked. It may be difficult for architects to effectively identify and correct potential problems when evaluating and verifying generated designs.

However, black box problems require a compromise between adaptability and manageability. Selecting the data set structure and fitting method can preserve the creativity of black box algorithms while reducing unreasonable results. Architectural design is a black box problem: "how to design" is the output of the black box process. High-quality data and standards for architectural design might reveal the black box of "how to design".

5.3 Limited Computational Infrastructure in Architecture

The complexity of architectural data far exceeds that of simple text or image data. It involves a large amount of geometric data, environmental parameters, and physical properties. Therefore, fully understanding and processing this data usually requires complex neural networks, and therefore has a higher demand for computing. However, existing algorithms and computing power often cannot meet the complex requirements of architectural design, and computing limitations have become a bottleneck for improving design capabilities.

Due to the rapid iteration of large models, computing limitations remain a key constraint on improving design capabilities. Training, fine-tuning, alignment, and reasoning of large models require a lot of computing. AI-assisted human–machine collaboration has great potential in improving design accuracy and efficiency, but requires continuous investment in computing and optimization. The construction industry urgently needs to establish a dedicated computing infrastructure and use external resources such as edge computing and cloud computing to ensure the efficient operation and sustainable development of large models.

5.4 Low Evaluability of Generated Results

Evaluating generated design solutions has become a key challenge. It is necessary to ensure that the generated designs are not only optimal in the short term, but also sustainable throughout the life cycle of the building. Most current generative designs focus on finding the best solution at the design stage, rather than considering a comprehensive assessment of long-term sustainability. For the sustainability assessment of a building's life cycle, its performance during construction, operation and maintenance, and demolition and material recycling must be considered.

At the same time, the environmental conditions of buildings are dynamic and are affected by factors such as climate change and urbanization, which can affect the long-term sustainability of buildings. Since environmental and economic conditions can change significantly at each stage, ensuring that the generated designs remain sustainable under these changing conditions increases the complexity of the evaluation process. Therefore, incorporating complex building environmental factors into the design generation stage, developing comprehensive life cycle sustainability assessment models, and enhancing dynamic environmental simulation capabilities are important foundations for evaluating and selecting generated content.

6 Innovative Directions in Generative AI-Assisted Architectural Design

6.1 Architectural Design Knowledge Graph

In the field of generative architectural design, building a specific representation framework and constructing a professional knowledge graph have become the key issues at the forefront of the discipline. Developing architectural knowledge graphs, clarifying data structures, selecting appropriate representation methods, and creating databases to centrally store, classify, manage, and retrieve knowledge are the key to GAI understanding architectural semantics and extracting architectural information. This framework will greatly improve the proficiency and accuracy of large models in tasks such as question answering, design generation, and discovery. The key to building an architectural knowledge graph is to effectively integrate data from multiple sources, including design drawings, documents, and market data. Automated knowledge extraction technologies such as natural language processing (NLP), text mining, machine learning, CNN, and multimodal models such as CLIP can be used to extract structured knowledge from large-scale unstructured data. Developing appropriate representation methods, such as early pixel- or voxel-based architectural morphology representation methods, and later graph-based architectural function relationship representation methods, can enhance the ability of artificial intelligence to understand architectural function semantics. It is also crucial to establish data standardization and standards, including standardized data formats, standardized terminology, and control data quality. By creating professional datasets with high-quality data sources, a solid foundation can be laid for building a knowledge base in the field of architecture.

6.2 Interpretability in Generative Design Processes

The customizability and interpretability of architectural design processes have become a top priority for the development of generative AI in the field of architecture. Unlike traditional rule-based generative methods, generative AI offers a unique approach. Due to its "black box" nature and data-driven characteristics, different design stages are often isolated and lack interconnection. By establishing a human–computer interaction framework tailored specifically for architects, the design of AI systems should be driven by the needs of architects and have the flexibility to adapt to the changing context of the architectural environment. Therefore, building a systematic generative AI-enhanced design process that integrates various methods and establishes an interpretable, adjustable and traceable architectural design workflow is essential to advance the knowledge production and creative paradigm of the discipline.

6.3 Integration of Generative Design and Construction

In the future, architectural generative design will aim to connect design and construction. Generative AI will no longer be limited to the generation of images, but will enable the transition from digital models to actual construction by integrating technologies such as 3D generation and BIM. This integration will improve the overall efficiency and quality of the project, and will also enable the construction industry to move from digitalization to intelligence, bringing new design possibilities from the conceptualization to the construction stage. Increasing digital collaboration between design and construction is essential. By combining creative design with technological strength, a direct transition from digital to physical buildings can be achieved. Although generative design is often associated with innovative structures and forms, some designs may be difficult to execute in the actual construction process due to the limitations of existing processes and methods. In addition, the complex geometries and structures created by AI usually require accurate digital models to represent. The process is controllable, for example, by creating a standardized combination of components. In addition, the key challenge is to effectively transform image-based generative design into a constructible model and use digital tools such as BIM for full process collaborative management.

7 Future Directions

By building professional knowledge maps, defining design processes, and integrating design and construction workflows, architecture will enter a new stage of intelligent empowerment, which will greatly enhance creative capabilities, strengthen comprehensive knowledge understanding, and improve architectural realization capabilities. In the era of artificial intelligence, it is crucial to establish a new knowledge framework with architects' creativity, critical thinking, and realization capabilities as the core. This framework will lay the foundation for sustainable development in the new era and give birth to a new model of human–machine collaboration.

Fund. Research fund of Tongji Architectural Design (Group) Co., Ltd. (Research on the technical system of solar residential units based on carbon reduction design (2023 J-JB04)).

References

1. Carpo M (2023) A short but believable history of the digital turn in architecture. e-flux Architecture (2023) https://www.efflux.com/architecture/chronograms/528659/a-short-but-believable-history-ofthe-digital-turn-in-architecture
2. Gainetdinov A (2023) Medium. Diffusion Models vs GANs vs VAEs: comparison of deep generative models. https://pub.towardsai.net/diffusion-models-vs-gans-vs-vaes-comparison-of-deep-generative-models-67ab93e0d9ae
3. Ho J, Jain A, Abbeel P (2020) Denoising diffusion probabilistic models. Adv Neural Inf Process Syst 33:6840–6851
4. Goodfellow IJ, Pouget-Abadie J, Mirza M, Xu B, Warde-Farley D, Ozair S et al (2014) Generative adversarial nets. Adv Neural Inf Process Syst 27. https://proceedings.neurips.cc/paper_files/paper/2014/hash/f033ed80deb0234979a61f95710dbe25-Abstract.html
5. Kingma DP, Welling M (2013) Auto-encoding variational bayes. Banff, Canada. http://web2.cs.columbia.edu/~blei/fogm/2018F/materials/KingmaWelling2013.pdf
6. Tarkhan N, Szcześniak JT, Reinhart C (2024) Façade feature extraction for urban performance assessments: evaluating algorithm applicability across diverse building morphologies. Sustain Cities Soc 1(105):105280
7. Lu Y, Wei W, Li P, Zhong T, Nong Y, Shi X (2023) A deep learning method for building façade parsing utilizing improved SOLOv2 instance segmentation. Energy Build 15(295):113275
8. Dong S, Wang W, Li W, Zou K (2021) Vectorization of floor plans based on EdgeGAN. Information 12(5):206
9. Sun Y, Jiang L, Zheng H (2020) A machine learning method of predicting behavior vitality using open source data. In: 40th Annual ACADIA conference (ACADIA 2020): distributed proximities. Association for computer aided design in architecture (ACADIA), pp 160–8. https://scholars.cityu.edu.hk/en/publications/publication(138c29ed-e9e9-4d59-9eae-adc6abc0ba1b).html
10. Tarabishy S, Psarras S, Kosicki M, Tsigkari M (2020) Deep learning surrogate models for spatial and visual connectivity. Int J Archit Comput 18(1):53–66
11. He Q, Li Z, Gao W, Chen H, Wu X, Cheng X et al (2021) Predictive models for daylight performance of general floorplans based on CNN and GAN: a proof-of-concept study. Build Environ 206:108346
12. Li X, Yuan Y, Liu G, Han Z, Stouffs R (2024) A predictive model for daylight performance based on multimodal generative adversarial networks at the early design stage. Energy Build 15(305):113876
13. Chokwitthaya C, Zhu Y, Mukhopadhyay S, Collier E (2020) Augmenting building performance predictions during design using generative adversarial networks and immersive virtual environments. Autom Constr 1(119):103350
14. Himmelb(l)au C. Coop Himmelb(l)au. [cited 2024 Aug 14]. Deep Himmelblau. https://coop-himmelblau.at/method/deep-himmelblau/
15. ZHA developing "most" projects using AI images says Patrik Schumacher. https://www.dezeen.com/2023/04/26/zaha-hadid-architects-patrik-schumacher-ai-dalle-midjourney/
16. Ibrahim NN (2022) AmazingArchitecture. AI-generated future cities by Manas Bhatia. https://amazingarchitecture.com/futuristic/ai-generated-future-cities-by-manas-bhatia
17. Portal to undiscovered atlantis by Karis|Futuristic. https://amazingarchitecture.com/futuristic/portal-to-undiscovered-atlantis-by-karisma-shoker
18. Alvarez L (2024) AmazingArchitecture. Chubineh Stadium by green clay architecture. https://amazingarchitecture.com/futuristic/chubineh-stadium-by-green-clay-architecture
19. Chang KH, Cheng CY, Luo J, Murata S, Nourbakhsh M, Tsuji Y (2021) Building-GAN: graph-conditioned architectural volumetric design generation. In: Proceedings of the

IEEE/CVF international conference on computer vision, pp. 11956–65. http://openaccess.thecvf.com/content/ICCV2021/html/Chang_Building-GAN_Graph-Conditioned_Architectural_Volumetric_Design_Generation_ICCV_2021_paper.html
20. Yang B, Li L, Song C, Jiang Z, Ling Y (2021) Automatic interior layout with user-specified furniture. Comput Graph 1(94):124–131
21. Sharma D et al (2019) A novel feature transform framework using deep neural network for multimodal floor plan retrieval. Int J Doc Anal Recognit (IJDAR). https://doi.org/10.1007/s10032-019-00340-1
22. Huang W, Zheng H (2018) Architectural drawings recognition and generation through machine learning. In: Proceedings of the 38th annual conference of the association for computer aided design in architecture, Mexico City, Mexico, pp 18–20. https://papers.cumincad.org/data/works/att/acadia18_156.pdf
23. Chaillou S (2020) ArchiGAN: Artificial Intelligence x Architecture. In: Yuan PF, Xie M, Leach N, Yao J, Wang X (eds) Architectural intelligence. Singapore: Springer Nature Singapore, pp 117–27. https://doi.org/10.1007/978-981-15-6568-7_8
24. Nauata N, Hosseini S, Chang KH, Chu H, Cheng CY, Furukawa Y (2021) House-gan++: generative adversarial layout refinement networks. arXiv:210302574
25. Shabani MA, Hosseini S, Furukawa Y (2023) Housediffusion: vector floorplan generation via a diffusion model with discrete and continuous denoising. In: Proceedings of the IEEE/CVF conference on computer vision and pattern recognition, pp 5466–75. http://openaccess.thecvf.com/content/CVPR2023/html/Shabani_HouseDiffusion_Vector_Floorplan_Generation_via_a_Diffusion_Model_With_Discrete_CVPR_2023_paper.html
26. Ali AK, Lee OJ (2023) Facade style mixing using artificial intelligence for urban infill. Architecture 3(2):258–269
27. Sun C, Zhou Y, Han Y (2022) Automatic generation of architecture facade for historical urban renovation using generative adversarial network. Build Environ 15(212):108781
28. Del Campo M (2022) Deep house—datasets, estrangement, and the problem of the new. ARIN 1(1):12
29. Gao W, Zhang X, He Q, Lin B, Huang W (2022) Command prediction based on early 3D modeling design logs by deep neural networks. Autom Constr 1(133):104026
30. Urbieta M, Urbieta M, Laborde T, Villarreal G, Rossi G (2023) Generating BIM model from structural and architectural plans using artificial intelligence. J Build Eng 1(78):107672
31. Del Campo M, Carlson A, Manninger S (2020) 3D graph convolutional neural networks in architecture design. In: Online and global, pp 688–96. http://papers.cumincad.org/cgi-bin/works/paper/acadia20_688
32. Bono G (2024) Text-to-building: experiments with AI-generated 3D geometry for building design and structure generation. ARIN 3(1):24
33. Horvath AS, Pouliou P (2024) AI for conceptual architecture: reflections on designing with text-to-text, text-to-image, and image-to-image generators. Front Arch Res 13(3):593–612
34. Zhang J, Liu N, Wang S (2021) Generative design and performance optimization of residential buildings based on parametric algorithm. Energy Build 244:111033
35. Zhang B, Mo Y, Li B, Wang Y, Zhang C, Shi J (2024) SIMForms: A web-based generative application fusing forms, metrics, and visuals for early-stage design. Singapore, pp 373–82. http://papers.cumincad.org/cgi-bin/works/paper/caadria2024_441
36. XKool Technology. https://www.xkool.ai/en
37. Zandavali BA, Jimenez Garcia M (2019) Automated brick pattern generator for robotic assembly using machine learning and images. In: Blucher design proceedings. Editora Blucher, pp 217–26. https://discovery.ucl.ac.uk/id/eprint/10117450/

A Cross-Modal AI Approach to Spatial Layout Generation: Fusing Graph, Text, and Boundary

Ximing Zhong[1](✉), Jiadong Liang[2], and Xianchuan Meng[3]

[1] Aalto University, Espoo, Finland
ximing.zhong@aalto.fi
[2] Architectural Association School of Architecture, London, UK
[3] Nanjing Univeristy, Nanjing, China
mxc@nju.edu.cn

Abstract. This paper introduces an innovative cross-modal spatial layout generation approach that leverages AI-agents to seamlessly integrate graphs, textual descriptions, and geometric boundary constraints for the creation of 3D room layouts. The proposed method utilizes a unique agent-based framework that incorporates large language models (LLM) and graph neural networks (GNN) to process and fuse multimodal inputs, allowing for a more comprehensive and flexible design process. By combining textual descriptions with boundary constraints into room feature embedding, the method enhances the semantic consistency and practicality of the generated layouts. Compared to the previous single-modal generative models, the experimental results demonstrate the method's effectiveness in accurately reconstructing room layouts and adapting to user-defined changes, showcasing its potential to revolutionize the field of architectural design by enabling the efficient integration of AI-agents and cross-modal data processing. Overall, this paper presents a significant step forward in the development of intelligent, adaptable, and user-centric tools for 3D spatial layout generation.

Keywords: Cross-modal · LLM · GNN · AI-agent · Spatial layout · 3D generation

1 Introduction

The three-dimensional (3D) spatial layout in architectural design is a complex task, due to the multiple objectives that influence the overall performance of the design outcome. Therefore, the spatial layout process in conventional architectural design is a dynamic, iterative, and time-consuming endeavour, constantly adjusted until the result with the best overall performance [1]. Location anchors (such as bubble diagrams), natural language descriptions of spatial attributes, and expected boundary of the design domain are extremely useful information in conventional design as essential reasoning information during the design process [2]. Fusing these different types of information is one of the challenges faced by AI based technology.

J. Fu et al. (eds.), *AI for Architecture*, Computational Design and Robotic Fabrication,
https://doi.org/10.1007/978-981-95-0974-4_2

2 Related Works

In the field of artificial intelligence, cross-modal typically refers to the ability of machine learning models to process and understand different types of data. This includes tasks like converting visual information (images) into text descriptions or transforming text into 3D models. Advances in cross-modal AI technologies like GANs, ChatGPT, and Codex [3] have opened possibilities for architectural design by incorporating other modal inputs that align better with conventional design practices. Cross-modal technologies can integrate textual descriptions, 2D images, and 3D data to make decisions for complex tasks and generate more accurate 3D models [4]. This cross-modal approach can provide more convenient, efficient, and time-saving innovative potential for room layout design, making it more in line with the habits of conventional design.

Text-to-3D generation is a crucial cross-modal technology in spatial generation methods. CG3D adopts a two-stage approach to generate 3D from text, involving structured relationship generation and physics-based optimization, aiming to address issues like text-controlled multi-object layouts, physical realism of scene compositions, and semantic and physical consistency [5]. MeshGPT generates triangle meshes directly as sequences of triangles, encoding triangle embeddings into latent quantized embeddings through a graph convolutional encoder and ResNet decoder, utilizing GPT to produce high-quality triangle sequences [6]. 3DMIT enhances LLMs' understanding of 3D scenes through end-to-end fine-tuning, injecting 3D information directly into LLMs and utilizing a 3D perceiver architecture with multiple encoders to extract and align global and detailed features in 3D scenes [7]. GALA3D utilizes large language models to guide initial layouts, then incorporates conditional diffusion priors to obtain more realistic and accurate real-world spatial layouts, ensuring spatial distribution and geometric consistency through adaptive geometry control and optimizing multi-object interactions in 3D scenes [8]. Text2Room proposes a method to generate complete, texture-rich 3D indoor scene meshes from textual inputs, first creating indoor layouts and furniture, then filling in gaps with 3D geometry, ensuring seamless and natural textures through depth alignment and mesh fusion [9]. However, these methods primarily record 3D models as meshes, with rendering as their primary task, lacking spatial topological relationships necessary for dynamic spatial layout design processes and adapting to architects' flexible design adjustments.

The topological relationships of subspaces are crucial for spatial plan generation methods. Data-driven approaches have initially demonstrated the effectiveness and controllability of architectural space generation. House-GAN++ is an intelligent framework for generating and optimizing floor plan layouts, representing room layouts as graphs, with nodes representing rooms and edges representing relationships between them, generating layouts based on input conditions like room count, types, and areas [10]. Zheng and Petzold decouple topological and geometric predictions of room layouts, using subgraph neural networks to predict room topologies and combining neural-guided plan sketching to generate layouts with reasonable topologies [11]. However, these methods focus primarily on data-driven generation from graphs to graphs, lacking other modal inputs, limiting the flexibility of on-demand adjustments during room generation. Spatial topological graph relationships overly abstract spatial requirements, whereas other modalities, such as boundary constraints or texts can more precisely store expected detailed information necessary for generation.

Boundary constraints are a type of geometric modality input that represents the design domain desired by the designer. Graph neural networks (GNN) can effectively integrate boundary constraints with topological constraints. Graph2Plan integrates user-in-the-loop design constraints based on given boundaries and bubble graphs. The system incorporates a retrieve-and-adjust paradigm, utilizes a deep neural network for floorplan generation, and includes a post-processing step for aligning room boundaries and vectorizing the floorplan [12]. Building-GNN emphasizes the controllability of the generative process that meets boundary conditions. Designers can traditionally define volumetric boundaries for the GNN model to infer volumetric details in space. Additionally, this method provides a flexible interactive design interface, realized through the GH-Python framework on the Rhino modelling platform, enabling architects to engage in a feedback loop with AI [13]. Utilizing building outlines and specified points as inputs, Graph-BIM is introduced, which encodes BIM data into graph structures tailored for AI learning, and helps AI understand and learn BIM data, enabling the generation of controllable 3D building models that meet spatial boundary constraints [14]. The results indicate that GNN models excel at generating vector models. Thus, we also adopted the previously advantageous GNN generation framework to explore architectural generation.

However, a common challenge with these methods is the need for human understanding to convert requirements into input features, as AI is unable to transform requirements into the features of input models across different modalities. For instance, in Building-GNN, architects cannot input text to modify the desired outcomes. In contrast, AI-agent technologies like the framework proposed by Building-Agent can intelligently convert textual requirements into 3D structured data, thereby reasoning to generate 3D layouts [15]. This enables us to use agents for data-driven feature engineering to inspire further development.

We propose a cross-modal spatial layout generation method that takes graph diagrams, such as bubble diagrams, and textual descriptions as inputs. This method utilizes an LLM-based agent framework and GNN to generate semantic 3D spatial layouts that meet boundary requirements and enables textual input to adjust room layout details during the generation process (Fig. 2.1). The way AI-agent integrates different modal information resembles the flexible process of conventional room layout design.

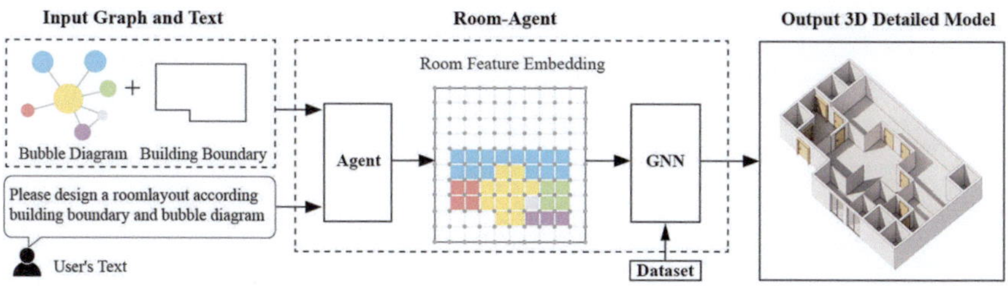

Fig. 2.1 Application of room-agent in architectural workflow

The contribution of this study lies in a cross-modal feature fusion method: utilizing various modalities such as spatial topological maps of mouse clicks, constraint boundaries, and text as input objects. Through the intelligent integration enabled by the AI-agent framework, this approach allows data-driven models to generate detailed 3D models that meet the dynamic evaluation needs of architects while conforming to the intricate layout features of datasets. Compared to earlier single-modal AI models like House-GAN, the inclusion of textual input enhances the model's ability for dynamic intelligent generation. Furthermore, the control through boundary constraints and graph anchoring ensures its potential for practical application in real projects.

3 Methodology

As shown in Fig. 2.2, the framework of the Room-Agent method is divided into three parts: (1) input representation, (2) Agent-driven feature engineering, and (3) GNN generation model. We initially employ graphs, text, and 3D boundaries as inputs, followed by utilizing multiple AI agents to facilitate the feature engineering process between the input and the GNN generation model. Specifically, the encoding unifies the requirements of different modalities into the features of graph nodes and edges. Finally, these features are fed into a trained GNN model, outputting the node and edge features of the graph, which are transformed into a new three-dimensional room layout. Additionally, text, anchor points, and boundaries can be used concurrently in the generation process for flexible adjustments to the layout, constituting an end-to-end modification process.

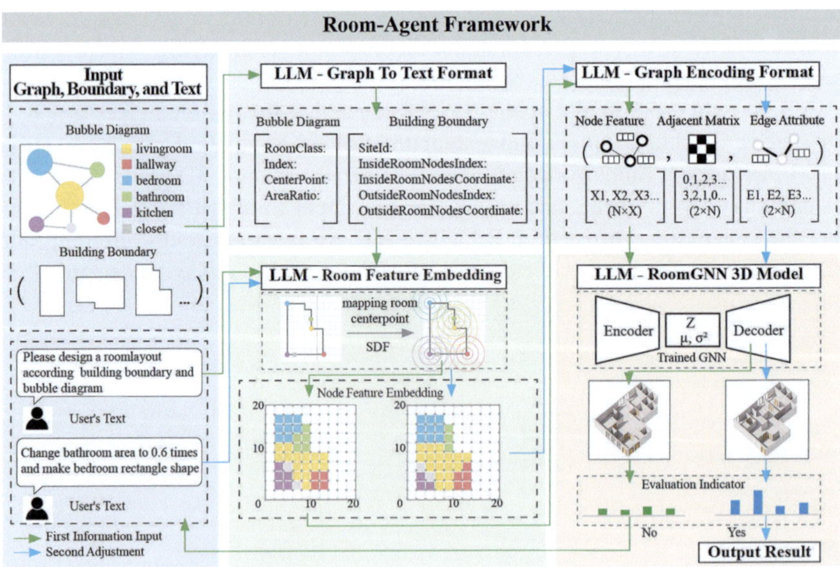

Fig. 2.2 Room-agent framework of converting graph, boundary, and text to 3D detailed model

In our approach, the bubble diagram and building boundary are transformed into text using AI agents that automatically reads and processes data formats to meet the requirements of structured data (Fig. 2.3). The graph and boundary vector information are converted into structured text output for accurate processing and efficient use of diverse data by another specialized agent that handles graph and boundary data. The "bubble diagram to text" agent function demonstrates how to convert a bubble diagram that represents rooms with coordinates and areas into a text format, while the "building boundary to text" agent function illustrates the method of translating building boundaries into text. Based on these structured data, room features first maps the coordinates of different room types within the building boundary, then colours the potential diffusion areas of functional rooms using Signed Distance Function (SDF), overlaying the diffusion area with a grid for evaluation, and calculates the influence probability of graph and text inputs on the properties of the final layout nodes, achieving feature embedding as input for AI generative models. By integrating features from different modalities into unified embedded features, AI agents provide more comprehensive and accurate inputs for generative models.

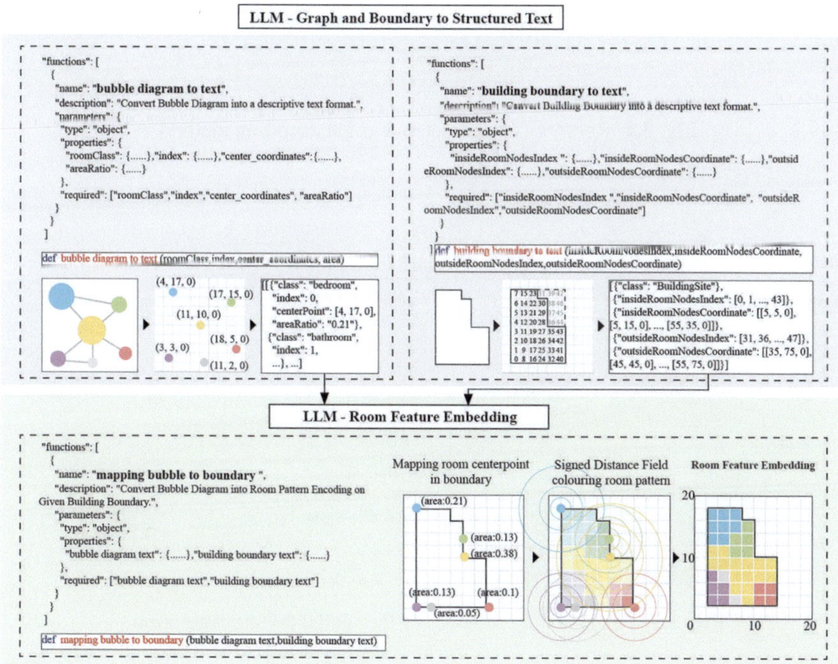

Fig. 2.3 LLM functions converting graph, boundary, and text to room feature embedding

GNN is employed as a generative model in this approach. Figure 2.4 demonstrates the details of the training architecture and generative process of RoomGNN. Figure 2.5 shows how graphs represent the details of 3D models. The model training process inputs node data encoded based on room feature embedding, while the model input consists of edge attributes. Loss is calculated by comparing the predicted results with the edge categories in the dataset layout, enabling the GNN model to learn the features between node inputs and spatial layouts, thus predicting the details of the spatial layout, as shown in the input and output sections of Fig. 2.5. Once the data training is completed, the feature embedding derived from user input is fed into the trained model to predict models. From the test dataset, we can observe the model's generalization and generative capability. Since we have set corresponding 3D components to different edge types, we can directly obtain the 3D model of the room layout.

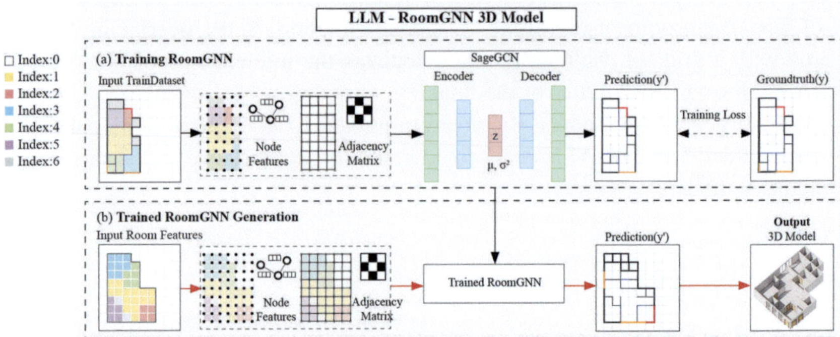

Fig. 2.4 Room-GNN dataset and graph encoding method

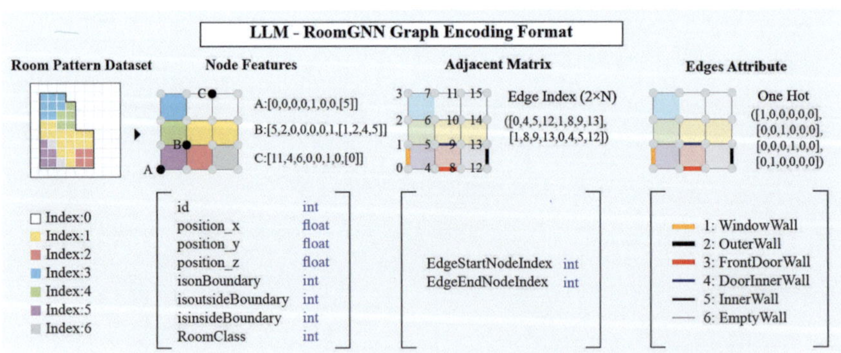

Fig. 2.5 The encoding method for the spatial layout generation

Figure 2.5 illustrates the details of how a 3D model is encoded into a graph data structure that AI can understand. In simple terms, nodes represent spatial grids, while edges indicate the categories of walls. The node features include properties such as coordinates, boundary types, and room categories. Edges are represented by an adjacency matrix defined by endpoints, and edge types are encoded using a one-hot encoding scheme to represent various wall types. This encoding method has been shown in previous studies to be superior to other encoding techniques for improving AI predictions of layout details [16]. During training, edge types are used for cross-entropy loss calculations, where the loss function reflects the discrepancies in spatial layout details. Therefore, a reduction in loss directly indicates effective learning.

As shown in Fig. 2.6, the AI model of RoomGNN is trained using the Housegan-Grid dataset, which is a modified version of the Housegan dataset. This dataset samples walls within a composite grid layout, offering precise delineation of room functions and various wall construction types.

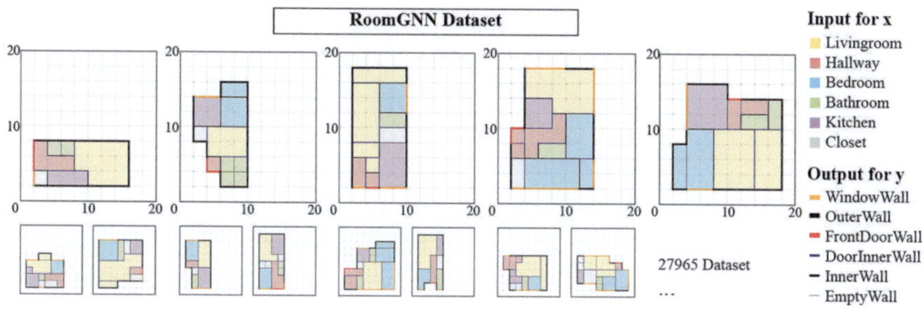

Fig. 2.6 Dataset used in room-agent

4 Experiment and Result

The results of RoomGNN on the test set, compared to the ground truth, demonstrate excellent learning capability, enabling the inference of complete room layouts from simplified room divisions. The average loss for the model's graph reconstruction results on the test set is 0.41 (Fig. 2.7b), as the loss itself represents the deviation of each layout-generating element from the dataset, showing its ability to effectively capture features related to spatial layouts within the dataset. The loss values of the generated results in Fig. 2.7a (1), (3), (4), (8), (9), (11) are significantly lower than the average, with the minimum loss at 0.207, indicating an almost complete reconstruction of the original spatial layout. This showcases RoomGNN's outstanding reasoning ability in translating room functions into detailed spatial layouts. However, when faced with complex architectural boundaries, the loss of RoomGNN's reconstructed spatial layouts tends to be higher. Notably, in Fig. 2.7a (2), (5), (6), (10), (12), the maximum loss reaches 0.565. Some reconstruction results exhibit additional walls both inside and outside the architectural boundaries, affecting the usability of the rooms, and some windowed exterior walls were not accurately predicted. Overall, RoomGNN's reconstruction performance on the test set highlights the effectiveness of this room-encoding approach, successfully learning layout logics such as wall types and spatial divisions.

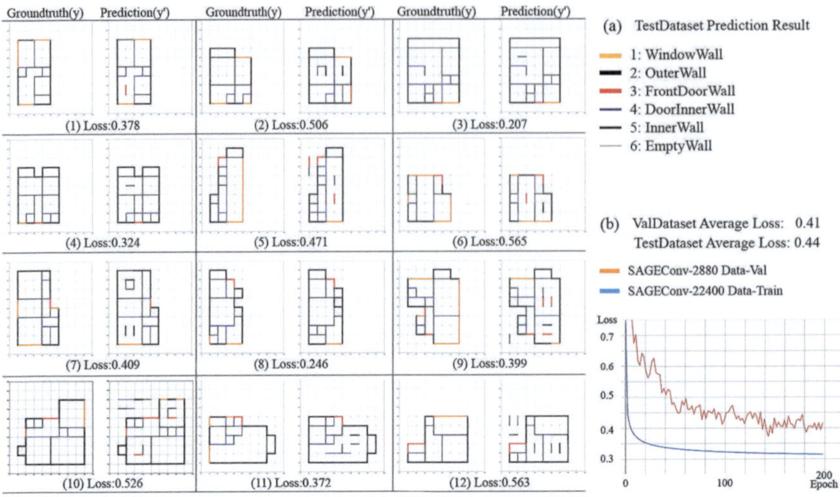

Fig. 2.7 Evaluation of room GNN graph reconstruction result based on the test dataset

Figure 2.8 demonstrates the generated room layouts of RoomAgent from various functional bubble diagrams and building boundary constraints. We analyzed the results using two metrics: Room Accessibility Index (RAI) and Room Enclosure Index (REI). In the generated results, the RAI is above 50% for all cases, indicating high accessibility. The first bubble diagram mapped to three building boundary layouts shows the highest room accessibility, with Fig. 2.8(1) reaching 100%. In terms of the REI, the spatial layouts show ideal room enclosure. For the third bubble diagram's generated results, Fig. 2.8(3) and (6) have an REI of 100%, indicating efficient room circulation. However, in Fig. 2.8(3) and (6), some wall types were not well predicted, such as multiple front doors in the entrance hall space. The second bubble diagram's overall generated results have lower RAI and REI indices. Figure 2.8(5) and (8) have an RAI of 57%, with nearly half of the rooms lacking door connections. Figure 2.8(8) has the lowest REI at 43%, where two living rooms lack outer boundaries and internal partitions, and the kitchen and storage room have become a single space without wall divisions. Overall, RoomAgent can effectively generate room layouts with detailed spatial divisions and high accessibility based on the provided functional bubble diagrams and building boundary constraints. However, there exists space for improvement in predicting wall types and handling layouts for spaces with special functions. These findings provide direction for further optimization of Room-Agent.

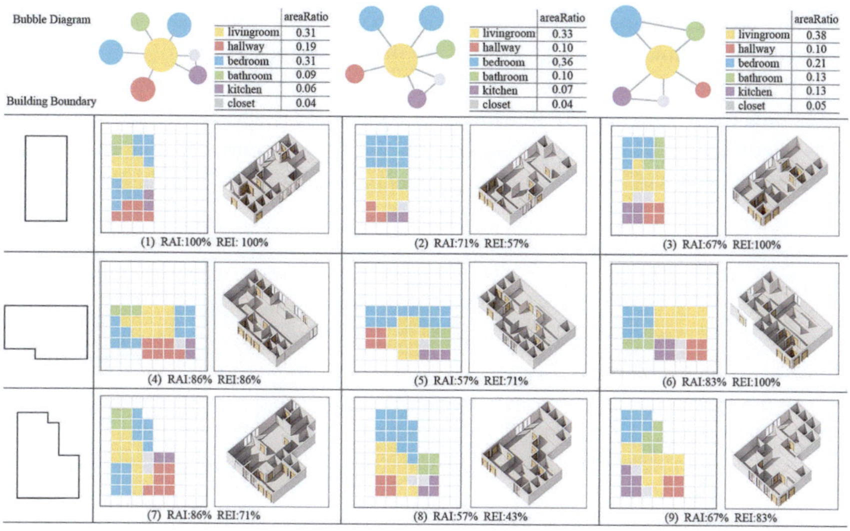

Fig. 2.8 Evaluation of room-agent generated result based on different bubble diagram and building boundary

Figure 2.9 shows the experimental results of modifying the spatial size and layout relationships through textual input. In the first modification, RoomAgent successfully met the user's request to reduce the bathroom area and optimize the bedroom into a rectangular layout. The modified three-dimensional spatial layout exhibited a high room enclosure rate and accessibility. In the second modification, Room-Agent repositioned the kitchen as requested, with the closet occupying the rectangular space in the lower left corner. The shape of each room in the current design is almost square, except shapes of the bedroom and bathroom. The LLM within RoomAgent, functioning as a probabilistic model, generated new layout results causing the bedroom's shape to become irregular, meaning a lack of precise control. These results suggest that RoomAgent can comprehend textual requirements and translate them into spatial layout outcomes, while also introducing some additional variations in the layout results due to the probabilistic nature of the large language model.

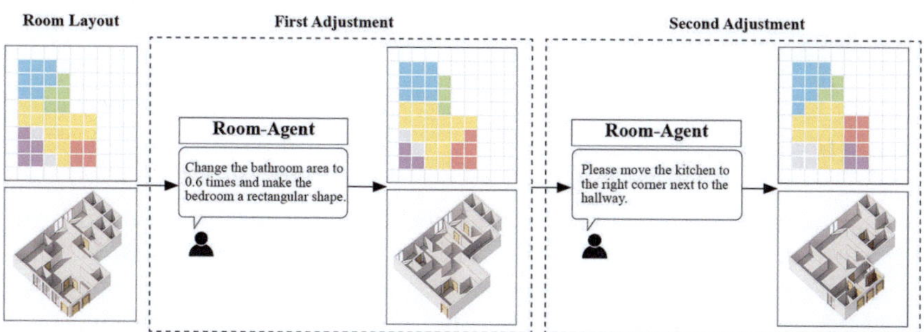

Fig. 2.9 Refining the room layout according user's requirement through natural language input

Compared to existing AI models, Room-Agent can define specific rooms and wall partitions within those rooms based on the constraints of the building boundary (Fig. 2.10). In contrast, previous generative AI models relied solely on abstract functional bubble diagrams, making it difficult to control the shape of the generated spatial layouts accurately. Room-Agent possesses the ability to accurately control the spatial layout generation based on the bubble diagram and the building boundaries.

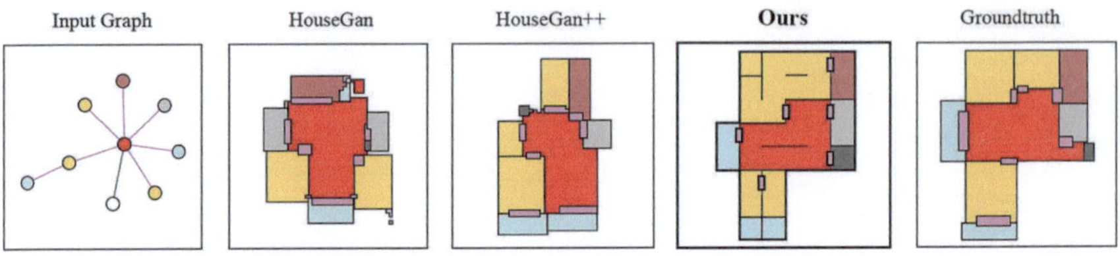

Fig. 2.10 Comparison of results between room-agent and existing AI models

5 Limitation and Future Work

Spatial layouts generated by the Room-Agent are constrained by the precision and diversity of the dataset. Experimental results indicate that some rooms fail to be with enclosed spaces, suggesting that the node features defining the outer boundaries need further refinement and diversification for training. Therefore, future improvements should focus on enhancing the accuracy and increasing the diversity of the dataset while providing more detailed features of each room. Compared to traditional parameter control methods, using textual input to control architectural layouts demonstrates clear advantages in efficiency and simplicity. However, we also observe that there may be instability in the outputs of large language models when generating spatial layouts. This could lead the model to modify room layouts without fully understanding the user's needs, thus affecting the output. Introducing more functional components within LLM is a suggested method to address this issue. The rule-based functions and the trained GNN model embedded in RoomAgent have already shown their diverse possibilities. To further enhance its ability to handle complex tasks, future work could consider using RoomAgent as a backbone, connecting various AI models across modalities, including video and voice inputs. Thus, RoomAgent could not only manage the design of architectural spatial layouts but also provide users with a more comprehensive and personalized design service through the integration of multimodal data.

6 Conclusions

This study proposes a novel cross-modal 3D room layout generation method aimed at addressing the limitations of traditional and current AI-driven approaches. By integrating graph-based representations (such as bubble diagrams) and textual descriptions into an LLM-based agent framework, combined with GNN, our method can dynamically and flexibly generate 3D spatial layouts that meet architectural design requirements. The incorporation of text input not only enhances the model's ability to generate intelligent and semantically consistent layouts but also allows for real-time adjustments during the generation process as needed. Boundary constraints further ensure the practicality of the generated layouts in real-world applications. Experimental results show that our method can accurately reconstruct room layouts, particularly in scenarios with clear spatial boundaries and functional divisions. Although challenges remain when handling complex boundary conditions, the Room-Agent system demonstrates potential in adapting to a wide range of design requirements through iterative refinement based on user input. The successful application of cross-modal (text, graphics, and spatial data) integration highlights the potential of our approach to bridge the gap between AI-generated designs and conventional architectural practices, providing valuable insights for developing more efficient, flexible, and designer-friendly tools in the field of architectural design.

References

1. Aalaei M, Saadi M, Rahbar M, Ekhlassi A (2023) Architectural layout generation using a graph-constrained conditional generative adversarial network (GAN). Autom Constr 155:105053
2. Emmons P (1998) The cosmogony of bubble diagrams. In: 86th ACSA annual meeting and technology conference, constructing identity, pp 420–425. Cleveland, Ohio
3. Cao Y, Li S, Liu Y, Yan Z, Dai Y, Yu PS, Sun L (2023) A comprehensive survey of AI-generated content (AIGC): a history of generative AI from GAN to ChatGPT. arXiv:230304226
4. Tan W, Ding Z, Zhang W, Li B, Zhou B, Yue J, Xia H, Jiang J, Zheng L, Xu X (2024) Towards general computer control: a multimodal agent for red dead redemption ii as a case study. arXiv:240303186
5. Vilesov A, Chari P, Kadambi A (2023) CG3D: compositional generation for Text-to-3D via gaussian splatting. arXiv:231117907
6. Siddiqui Y, Alliegro A, Artemov A, Tommasi T, Sirigatti D, Rosov V, Dai A, Nießner M (2024) MeshGPT: generating triangle meshes with decoder-only transformers. In: Proceedings of the IEEE/CVF conference on computer vision and pattern recognition, pp 19615–19625
7. Li Z, Zhang C, Wang X, Ren R, Xu Y, Ma R, Liu X (2024) 3DMIT: 3D multi-modal instruction tuning for scene understanding. arXiv:240103201
8. Zhou X, Ran X, Xiong Y, He J, Lin Z, Wang Y, Sun D, Yang M-H (2024) GALA3D: Towards text-to-3D complex scene generation via layout-guided generative gaussian splatting. arXiv: 240207207
9. Höllein L, Cao A, Owens A, Johnson J, Nießner M (2023) Text2Room: extracting textured 3D meshes from 2D text-to-image models. In: Proceedings of the IEEE/CVF international conference on computer vision, pp 7909–7920
10. Nauata N, Hosseini S, Chang K-H, Chu H, Cheng C-Y, Furukawa Y (2021) House-gan++: generative adversarial layout refinement network towards intelligent computational agent for professional architects. In: Proceedings of the IEEE/CVF conference on computer vision and pattern recognition, pp 13632–13641
11. Zheng Z, Petzold F (2023) Neural-guided room layout generation with bubble diagram constraints. Autom Constr 154:104962
12. Hu R, Huang Z, Tang Y, Van Kaick O, Zhang H, Huang H (2020) Graph2plan: Learning floorplan generation from layout graphs. ACM Transactions on Graphics (TOG) 39:118: 111–118: 114

13. Zhong X, Koh I, Fricker P (2023) Building-GNN: Exploring a co-design framework for generating controllable 3D building prototypes by graph and recurrent neural networks. In: International conference on education and research in computer aided architectural design in Europe. eCAADe, pp 431–440
14. Liang J, Zhong X, Koh I (2024) Bridging BIM and AI: a graph-BIM encoding approach for detailed 3D layout generation using variational graph autoencoder
15. Zhong X, Liang J, Li Y (2024) Building-agent: a 3D generation agent framework integrating large language models and graph-based 3D generation model. In: International conference on education and research in computer aided architectural design in Europe, Nicosia, Cyprus. eCAADe
16. Liang J, Zhong X, Koh I (2024) Bridging BIM and AI: a graph-BIM encoding approach for detailed 3D layout generation using variational graph autoencoder. In: International conference on computer-aided architectural design research in Asia. Association for Computer-Aided Architectural Design Research in Asia, pp 221–230

Discussion on the Ontology Inconsistency and 3D Model Generation in Architectural Design Assisted by AIGC

Yunfan Ye, Fan Wang, Haojia Xu, and Hui Wang(✉)

College of Civil Engineering and Architecture, Zhejiang University, Hangzhou, China
wang_hui@zju.edu.cn

Abstract. AIGC technology is playing an increasingly crucial role in generative design. At present, many AIGC tools are used to generate images in the architectural design, such as Midjourney, Stable Diffusion. Although the single image generated by AIGC tools might have high quality, it is difficult to maintain the consistency of the ontology when representing the same building from different angles due to the inevitable randomness of the generation process, which means different pictures contradict each other in details. This article discusses the possible solutions of ontology inconsistency problem with the guidance of different AIGC tools, and points out that generating 3D model is the most effective way to solve the problem, but the current generation tools still have various defects. With the contrast of different approaches, the technical path of Sketch-Multiple Pictures-Contradiction Elimination-3D Model is considered to be more appropriate for architectural design process. This path also puts forward technical requirements for the future specialization of current AIGC tools.

Keywords: Ontology inconsistency · AIGC · Architectural design · 3D model generation

1 Introduction

The rapid development of Artificial Intelligence Generated Content (AIGC) has brought new patterns of design work and also driven a significant reform in the field of architectural design. AIGC refers to the use of generative AI algorithms to assist or replace humans in creating rich personalized and high-quality content, such as text, painting, music, video, and even interactive 3D content, at a faster pace and lower cost, according to user input and requirements [1]. Through a large number of model training, AI understands the patterns and rules, and generates some new content with similar characteristics to the training data. Presently, AI is able to process and generate text, images, videos and other modes of information, and obtain a large number of outputs in a short period of time, which greatly improves design efficiency.

The research on AIGC in architectural design field is in the ascendant. Yanfeng Huang et al. summarized the current three generation methods of AIGC, and proposed

J. Fu et al. (eds.), *AI for Architecture*, Computational Design and Robotic Fabrication,
https://doi.org/10.1007/978-981-95-0974-4_3

an optimization method for the full generation of architectural design: generating 3D models from 2D images and prompts, and then slicing 3D models into 2D slices to generate specific planes [2]. Guo Li et al. generalized the main application scenarios of generative AI in architectural design fields, developed by various design institutes and universities [3]. Taking *LookX AI Cloud* as an example, Wanyu He et al. discussed how to integrate generative AI technology and proposed an intelligent workflow that integrates language model to realize conversational interaction design [4]. Chao Yuan explored the multi-dimensional generation mode of AI and put forward a new architectural design mode under the influence of generative AI [5].

This article focuses on the problem of ontology inconsistency in term of the deep use of AIGC in architectural design, which aims at a throughout generation of complete design scheme rather than generating a single perspective image. The problem will be demonstrated firstly, and then the possible solutions within the image generation tools will be explored. Aiming at solving the problem completely in the future, different tools of generating 3D models and the feasible working path in architectural design is also to be discussed.

2 Ontology Inconsistency

At present, the architectural AIGC mostly focuses on generating a certain style of renderings through sketches (including hand-drawn drawings or wireframes of simple models), which can be compared with a variety of styles in the early research stage and inspire design ideas. Tools such as Midjourney, Stable Diffusion and Dall-E are commonly used in this approach. However, the further use of AIGC tools will inevitably encounter the problem of ontology inconsistency. It refers to the generated results cannot maintain the consistency of forms and details, when generating different perspectives of the same building. Adding more manual details to control the model will reduce the proportion of AI-generated design, and is equivalent to reducing the AIGC tool to a renderer. This problem has a great impact on the deep use of AI in architectural design.

For example, in Stable Diffusion, the checkpoint and LoRA of intended design style should be firstly determined before input the corresponding positive and negative prompts. Then the outline of one perspective of a building should be imported into *ControlNet* under the control of *Lineart*. Next, different parameters, like the iterative steps and control weights should be adjusted to generate multiple renderings of that perspective. After that, the outlines of other perspectives could be imported into *ControlNet* for wireframe control, to generate other renderings, with the mostly consistent parameters and prompts. It is the results of four different perspectives of a building model (Fig. 3.1). As the checkpoint, LoRA and the basic parameters remain unchanged, the generated renderings have the similar style and surroundings. However, there are still obvious problems such as inconsistencies and contradictions within the volume, facades and material details, which means the ontology is not the same one.

Fig. 3.1 Inconsistencies in generating multi-angle perspectives based on wireframes (Drawn by the authors)

The fundamental cause of the problem lies in the deep generative mechanism of AIGC tools. In the field of image generation, there are four major generation models in recent years: GAN, VAE, Flow based Model, Diffusion Model, etc., which are mainly based on deep learning as the training model. Starting in 2022, the main popular image generation model is Diffusion Model. For example, Stable Diffusion uses latent diffusion model. Historically, probabilistic models suffer from a trade-off between two conflicting objectives: tractability and flexibility. Diffusion models are inspired by non-equilibrium thermodynamics. It defines a Markov chain of diffusion steps to slowly add random noise to data and then learn to reverse the diffusion process to construct desired data samples from the noise. However, when random noise increases the diversity of the generated result, it also causes the inevitable randomness.

The issue of ontology inconsistency is currently a widespread problem in the field of AIGC, especially in the case of generating multiple images or videos of the same subject. How to maintain the consistency of the ontology with less manual control needs more exhaustive exploration. Developers of AIGC tools have taken note of this issue and have attempted to address it in certain ways, such as by adding reference images to enhance similarity. Recently Midjourney has released the character reference tool in version 6.0 to maintain consistent characters across multiple images. It can optimize the character traits such as face, hair and dress, to obtain a roughly consistent result. But it cannot achieve accurate sameness and is not efficient in architectural design.

3 Optimization Within Image Generation Tools

The possible solutions within 2D image generation tools can be explored. In Stable Diffusion, there is also a reference function similar to the character reference function in Midjournal. It can guide diffusion using single image as a reference without any control model. However, currently this function only has good diffusion results for characters, animals, objects and so on. For architecture, even with the control of line draft and depth of field, it can only generate images with similar styles, and there is a significant randomness in the details. The Seed parameter in Stable Diffusion determines the content of the generated image and ensures the reproducibility of the generated results. As long as the same number of seeds and other identical parameters are used, the same image result can be obtained. In the process of generating architectural renderings, modifying other parameters appropriately and using the same number of seeds can obtain other renderings with the same style features and details.

After research, it was found that the number of seeds can control the similarity of generated images, but to maximize the consistency of the generated result model ontology, more controls need to be added. For example, in Stable Diffusion, it is viable to optimize the generated contents by adjusting parameters and importing images circularly into *ControlNet*. The seed of the first image from the above generation process could be used to control the generation of the other perspectives. The generated images could be imported into *ControlNet* to add control by *Reference*, which is beneficial to generate similar images. Setting parameters and sending images to *inpaint* to repaint the details also contribute to the consistency of the pictures. If there are differences among the first one and other perspectives, the images from other perspectives could be used to add control by *Reference* with the same seed. The regenerated picture could develop in the way as above to improve the details.

Under multiple iterations by this method, the 2D generated images could improve the result of Fig. 3.1 to a certain extent, with the improved consistency of different perspectives (Fig. 3.2). However, there are still discrepancies in details. To achieve further consistency of 2D images, other tools need to be explored.

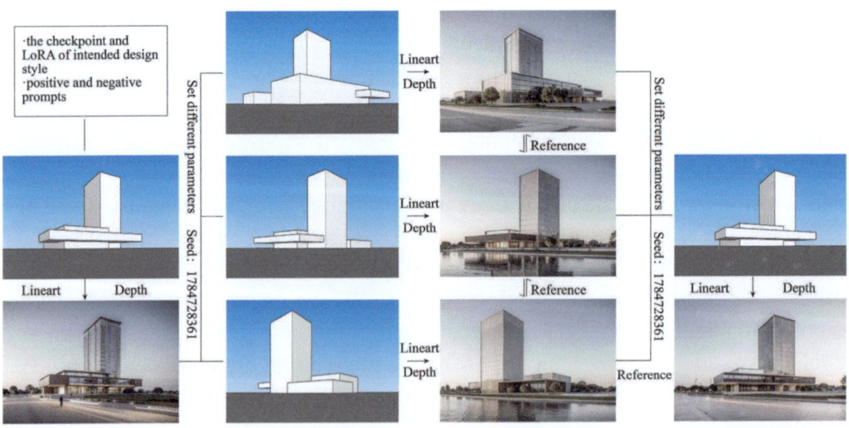

Fig. 3.2 The process of generating multi-perspective images with higher consistency in stable diffusion (Drawn by the authors)

4 Present Approaches of 3D Model Generation

Due to the inevitable randomness in the image generation process, it can hardly solve the problem of ontology inconsistency solely by relying on graphical AIGC tools. It can be inferred that generating a unified 3D model is the most effective and complete solution, which is also beneficial for interfacing with the subsequent architectural design process. At present, there are two common approaches of 3D generation: text-to-3D and image-to-3D, which are mostly used in the fields of game and animation generation.

4.1 Text-to-3D

Some AI tools for generating 3D models have emerged, such as Meshy, Masterpiece X and Genie (Table 3.1). These tools allow users to input texts (prompts) to generate 3D models, with output formats typically including *. obj, *. fbx, etc. These tools can generate models with textures in short time, and some can generate multiple models at once. In the early stages of architectural design, it may be useful for rough morphological research. However, as it main focuses on the field of game design, there are problems such as the models lack a sense of scale and rounded edges.

Table 3.1 Overview of different types of text-to-3D tools

Tool	Input	Output	Advantage	Disadvantage
Meshy	Prompts	3D model with maps (*.fbx/*.obj/*.glb/*.usdz/ *.stl/*.blend)	Quick generation process; high quality and modifiable model	Unsharp delineation; vague model components; non-editable model
Master piece X	Prompts	3D model with maps (*.obj)	Model with sharp corners and a strong sense of Volume; ability of importing into a virtual reality environment; clear components;	Combination of building clusters; glued-together blocks; poor quality model
Genie	Prompts	3D model with maps (*.fbx/*.obj/*.glb/*.usdz/ *.stl/*.blend)	Multiple generation models; iterative operation	Sticky components; poor editing capabilities

4.2 Image-to-3D

Generating 3D models from images is another possible solution. Such tools can be used to generate a model of the whole building by inputting images of the building facade. At present, several AI tools for generating 3D models have emerged (Table 3.2), and some of them also support the import of video.

Table 3.2 Overview of different types of image-to-3D tools

Tool	Input	Output	Advantage	Disadvantage
Wonder 3D	Single image	3D model with maps (*.ply)	Strong ability to recognize shapes	Unclear relationship between shapes; model with undulating plane and over-smooth edges; non-ideal generating effect
IMAGE to STL	Single image without background	3D model with maps (*.glb/ *.obj/ *.ply)	Fast model generation; rich output format; suitable mode for LOGO or terrain modelling	Inaccurate image recognition; non-ideal quality, detail and texture of model; model doped with blocks
Tripo AI	Text/single image	3D model with maps (*.glb/)	Strong ability to recognize planes and basic shapes; better quality of generation	No adjustable parameters; model of a single trilateral closed grid
Luma AI	A video with a stable scene; multiple pictures from different perspectives for a subject	3D model (Luma field format)	Outstanding ability to draw model and identify the form, especially the transparent components such as glass	Non-editable model; incompatible with architectural software

Compared with Text-to-3D, the approach of image-to-3D is closer to the working flow of architectural design. The architect could generate multiple images using prompts or sketches, and then use image-to-3D tools to generate building models. Videos generated by SORA could also be imported into Luma AI to generate the 3D model. However, due to the expression of complex bodies, like spot light, in Sora video is still not perfect, and the model is somewhat blurred.

Overall, the current AI tools for generating models based on text or image are not mature, but tools like Luma AI which can provide accurate models based on multiple images are expected to be applied in architectural design in the future.

5 Exploration of the Feasible AIGC Path for Generating Architectural Models

To generate 3D model is an ultimate object of AIGC-assisted architectural design. It requires at least the following four aspects: (1) generating results in line with the spatial composition logic of architectural design; (2) reducing manual intervention as much as possible; (3) having high generating efficiency; (4) the output files are suitable for the needs of the architectural design work in the subsequent stages. It is difficult to achieve the above objectives at the same time with the current technology level. But based on the work logic of architectural design, the future technical path of AI-assisted architectural design can be discussed.

5.1 Path 1: Text-to-3D Models

Text-to-3D path has an advantage in dealing with the consistency of multiple perspectives of single model. This path enables AI to modify model details in a holistic and continuous manner, effectively circumventing the limitations that 2D-to-3D tools only define one or a few views of the model. However, according to the currently available Text-to-3D tools, there are three insurmountable difficulties. Firstly, the poor quality of model generation for uncertainty cannot be applied with simple types. Secondly, it is difficult to accurately describe the composition of the building in a textual way. Thirdly, the results of generation that adhere to the functional logic and the myriad of design-specific rules is a formidable obstacle. Consequently, the feasibility of utilizing this path for architectural design in the immediate future appears tenuous.

5.2 Path 2: Text-Video-3D Models

Text-to-video tool is able to generate continuous video scenes, which has the potential to promote its application in architecture field. As an open-sourced software, Stable Video Diffusion 1.1 provides two input methods, prompt or image. Based on the prompt or image, high-quality video files can be obtained by adjusting camera parameters. The video or video-based multi-angle images can be input into softwares such as Luma AI for model generation.

However, generating a panoramic video of a building through text also faces the same problems in path 1: the building form is difficult to be described accurately, and the generated result is difficult to conform to the logical needs of the design. Furthermore, if the scene in the generated video is complex or contains contradictions, the existing tools cannot be used to generate 3D models. For example, it was found that a video of a seaside castle generated by Sora was invalid when imported into Luma AI.

5.3 Path 3: Sketch-Multiple Pictures-Contradiction Elimination-3D Model

Through the above analysis, it can be seen that the overall control of the architect is essential in the preliminary stage of design with AIGC tools. Generally a sketch drawn by architect is based on reasonable composition logic which determines the outline of the building according to the site and function requirements. It can give a direction of the AI generation while ensuring the rationality of the generated results.

According to the current technological level, the Image-to-Image mode cannot completely eliminate the contradictions between different pictures. A feasible solution is to optimize the generation process of the 3D model tools, such as specifying a prior picture or a prior part in one picture which will be adopted preferentially. The false content in other pictures will be ignored so that the consistency can be maintained.

It's considered that the path of Sketch-Multiple Pictures-Contradiction Elimination-3D Model is a possible approach to architectural model generation. This path puts forward certain technical requirements for the specialization of 3D model tools at present, and points out a feasible direction for the development of AI-assisted architectural design.

6 Conclusion

As a widespread problem in the field of current AIGC, ontology inconsistency poses a significant barrier to the in-depth application of AIGC in the field of architectural design. This article describes the problem and attempts to optimize the solution within the image generation tool, resulting in a limited improvement in the generated outcomes. To completely resolve this issue and integrate with the architectural design workflow, generating 3D models is considered as an effective solution. The article further analyses the characteristics and advantages and disadvantages of different 3D model generation tools, such as *text to 3D* and *image to 3D*, and discusses their potential for application in the field of architectural design in the future. This article considers that the path of Sketch-Multiple Pictures-Contradiction Elimination-3D Model is more appropriate for architectural design logic. It is important that additional step should be added to eliminate contradictions and the model generation tools should support local re-generation. This path provides a reference for the further development of AI-assisted architectural design. However, the tools mentioned in the article are currently under rapid evolution, so the feasible paths presented here are only phased cognition. It is expected that with the iteration of AIGC, highly integrated tools will emerge in the coming future and solve the current ontology inconsistency problems.

Acknowledgements. This research was supported by the National Natural Science Foundation of China (52378043).

References

1. Wang Y, Pan Y, Yan M et al (2023) A survey on ChatGPT: AI–generated contents, challenges, and solutions. IEEE Open J Comput Soc 4:280–302
2. Huang Y, Zhou Y (2023) Exploration on the generative architecture design method with AIGC technology: a case of the overall design process of generating architectural image with prompt as a key word. Urban Arch 15:202–206+213
3. Li G, Zhang T, Xing Z (2023) On the eve of the technological revolution: architectural and scene design innovation under the wave of generative AI tools. Chin Overseas Arch 9:24–28

4. He W, Yong LSJ (2023) Exploration of generative artificial intelligence in architectural design a case study of LookX AI cloud. Arch J 10:36–41
5. Yuan C, Zheng H (2023) A new architectural design methodology in the age of generative artificial intelligence. Arch J 10:29–35

Performance Prediction of AI-Generated Architectural Layout Design: Using Daylight Performance of Residential Floorplans as an Example

Xiao Hu[1], Hao Zheng[2(✉)], and Dayi Lai[1(✉)]

[1] Shanghai Jiao Tong University, Shanghai, China
{huxiao1009,dayi_lai}@sjtu.edu.cn
[2] City University of Hong Kong, Hongkong, China
hazheng@cityu.edu.hk

Abstract. The integration of artificial intelligence (AI) in architectural design, especially for generating floor plans, can greatly streamline the process. However, most AI-generated plans focus on form and spatial layout, often neglecting crucial performance evaluations because they are presented as images without the necessary geometric and physical properties for effective simulation. To address this limitation, we propose a novel approach that combines diffusion models with generative adversarial networks (GANs) for generating and evaluating floor plans. We fine-tuned a Low-Rank Adaptation (LoRA) model for creating residential floor plans, while a GAN quickly predicts daylighting performance. Our results show that the diffusion model generates a more varied set of floor plans compared to the training set. The GAN accurately assesses daylighting performance, with deviations from the ground truth not exceeding 5%, achieving a mean squared error (MSE) of 4.2 and a structural similarity index (SSIM) of 0.98. Additionally, it operates 267 times faster than traditional methods. This approach equips architects with a reliable tool for efficient early-stage design decisions, enhancing AI-driven workflows.

Keywords: Automated floor plan · Diffusion model · Generative design · Conditional generative adversarial network

1 Introduction

1.1 Research Background

The use of automated tools in architectural floor plan generation, especially for residential design, is gaining traction due to their ability to enhance efficiency by reducing repetitive tasks and minimizing trial and error in the design process [1]. Automated floor plan generation typically utilizes two main methods: rule-based and learning-based approaches [2]. Rule-based methods rely on algorithmic constraints derived from design principles, using predefined generative rules, shape grammar [3], and advanced techniques like

J. Fu et al. (eds.), *AI for Architecture*, Computational Design and Robotic Fabrication,
https://doi.org/10.1007/978-981-95-0974-4_4

agent-based modeling [4, 5] and graph theory [6, 7]. However, these often require architects to possess specialized skills in mathematics and programming, creating barriers to widespread adoption. In contrast, learning-based methods have gained traction due to their accessibility and ease of implementation, allowing architects to bypass complex algorithms by extracting features from existing data. The rise of machine learning and AI has led to data-driven approaches that learn from extensive design datasets, resulting in diverse and creative floor plans. Techniques like Generative Adversarial Networks (GANs) [8], supervised learning with Auto-Encoder refinement [9], and Graph Neural Networks (GNNs) [10] exemplify this trend, making it increasingly common for architects to integrate AI tools into their workflows for efficient floor plan generation.

Despite advancements in automating floor plan generation, the evaluation of the generated designs is still lacking, particularly in the context of learning-based approaches. While rule-based methods facilitate the straightforward definition and adjustment of design variables for performance optimization [11], learning-based methods often embed these variables within generated images, thereby complicating traditional evaluation techniques. This necessitates the use of specialized software for geometric modeling and simulation, which can be time-consuming and requires expertise. To address this gap, recent developments in deep learning, particularly with Generative Adversarial Networks (GANs) like pix2pix [12], show promise in converting generated images into actionable performance data.

Nevertheless, the incorporation of performance assessment into the domain of learning based architectural design has been largely neglected. This paper puts forth innovative model architectures and strategies that integrate generation with performance evaluation, leveraging diffusion models and conditional GANs (cGANs) to enhance the usability of AI-generated floor plans and equip architects with invaluable tools for early-stage design decisions.

1.2 Literature Review

Diffusion models are emerging as a powerful alternative to GANs in automated floor plan generation. While GANs, have been widely used for image generation, diffusion models like Denoising Diffusion Probabilistic Models (DDPMs) [13] and Latent Diffusion Models (LDMs) [14] offer enhanced stability and diversity. These models can leverage textual descriptions for guided image generation, expanding their applicability. Recent studies have demonstrated the potential of diffusion models in architectural design, with innovations like HouseDiffusion [15] effectively generating vector floor plans that maintain geometric relationships, and FloorplanDiffusion [16] allowing multi-conditional inputs to produce designs. However, a key limitation of these models is their neglect of environmental performance factors such as daylighting and ventilation, which are critical for practical applications in early-stage design.

In the realm of performance prediction, deep learning methods, particularly Convolutional Neural Networks (CNNs) and GANs, have streamlined the assessment of building performance, which is essential for creating efficient and sustainable environments. Traditional assessment methods can be complex and time-consuming, but DL offers efficient alternatives. For instance, multimodal GANs have reduced computational time for daylight prediction significantly [17], while CNNs have provided real-time feedback on daylight performance in floor plans [18]. Additionally, GANs have shown promise in Computational Fluid Dynamics (CFD) for urban airflow prediction and optimizing environmental factors [19]. Despite these advancements, integrating performance evaluation with AI-generated designs remains largely unexplored, presenting an opportunity for further research to enhance the efficiency of both design and assessment processes.

1.3 Goal of This Study

The goal of this study is to create a streamlined workflow that integrates performance optimization into AI-generated residential floor plans. This research addresses the complex relationships between design generation and environmental performance by combining the generative strengths of diffusion models with the performance evaluation capabilities of GANs. The main contribution is an automated system that generates diverse and functional floor plans while providing real-time feedback on daylight performance, facilitating rapid design iterations. By merging design generation with performance evaluation, this workflow enhances the practicality and effectiveness of AI-driven design in architectural practice, allowing for more informed decision-making.

2 Methodology

2.1 Overall Framework

The overall framework enables designers to quickly generate and evaluate architectural floor plans using a workflow that combines a diffusion model and a conditional GAN (cGAN), specifically the pix2pixHD [20] model. Architects start by creating floor plan layouts from text prompts with the diffusion model. These layouts are then analyzed by the cGAN to predict their performance. The process involves two main steps: first, a rule-based parametric model is developed using Rhino and Grasshopper to create floor plans that adhere to specific spatial requirements, generating a dataset for the diffusion model. Second, the Ladybug and Honeybee plug-ins perform simulations to produce daylight performance data, which, along with the floor plan images, forms the training dataset for the pix2pixHD model (Fig. 4.1).

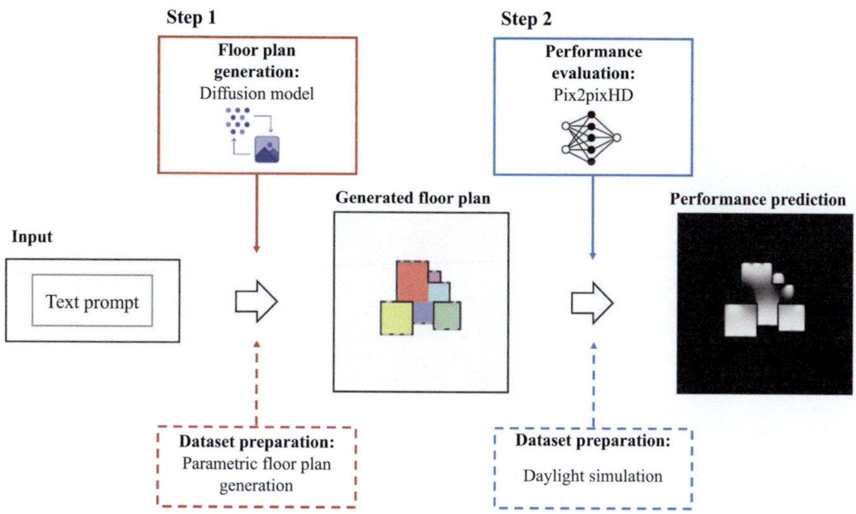

Fig. 4.1 Overview of the workflow

2.2 Floor Plan Generation

In this study, we use the diffusion model to generate building floor plans. We fine-tune the diffusion model using the Low-Rank Adaptation (LoRA) model [21], which is designed to adapt large pretrained models to new datasets without requiring full retraining from scratch. To prepare the dataset, we use a parametric algorithm to generate floor plans.

Dataset for Training LoRA

In the process of generating floor plans, a dataset is required for training. A parametric algorithm was developed to generate floor plans as a dataset. The method combines different rooms by applying forces between them and using the final balance of forces to create a tightly connected floor plan. As illustrated in Fig. 4.2, the creation of a residential floor plan entails considering the number of rooms, the area of each room, and their topological relationships. The process comprises two main phases: room generation and force application to nodes.

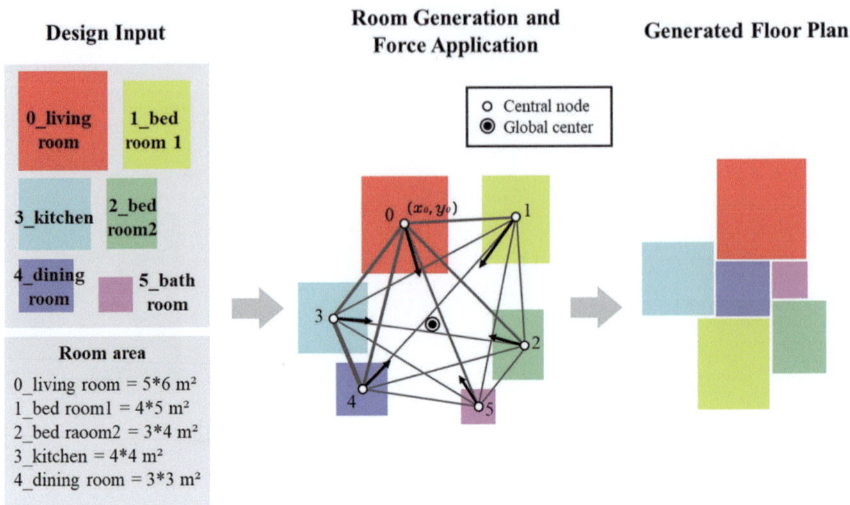

Fig. 4.2 Generating floorplans through parametric generation algorithm

Room generation phase: the coordinates (x_i, y_i) of the central node determine each room's initial position. Rectangular shapes are created with the central node as the center, representing each room's designated area.

Force application phase: forces are applied to nodes in two steps. First, a global center point (typically the origin) is established, and an attractive force clusters all room nodes around this center. Then, varying spring forces between room nodes ensure their topological relationships match the design inputs. For instance, dining rooms and kitchens typically have strong connections, so the spring forces between their nodes are set very high. If these rooms are not adjacent, the spring forces will attempt to move them closer together. The dataset generated by the script and used to train LoRA is shown in Table 4.1.

Table 4.1 Dataset used to train LoRA

Type	Data volume	Area/m^2	Cases from dataset
5_rooms	320	50	
6_rooms	320	100	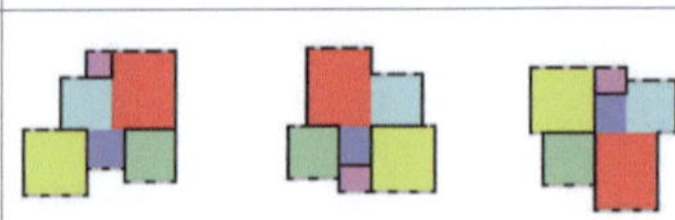
7_rooms	320	120	

(*continued*)

Table 4.1 (*continued*)

Type	Data volume	Area/m^2	Cases from dataset
Legend of room types		Living room (255,99,99); Kitchen (152,255,255); Dining room (152,158,255); Other room (220,0,255); Bed room1 (241,255,99); Bed room2 (152,255,165); Bath room (233,152,255)	

Diffusion Model and LoRA Training

Diffusion models work by adding Gaussian noise to training data (forward diffusion process) and then learning to reverse this process step by step to recover the original data (reverse diffusion process). Diffusion-based image generators have seen widespread commercial interest, such as Stable Diffusion and DALL-E. These models typically combine diffusion models with other models, such as text-encoders and cross-attention modules to allow text-conditioned generation. In this study, we use stable diffusion to finish our tasks.

Low-Rank Adaptation (LoRA) is trained to adapt the weights of Stable Diffusion to learn floor plan layout patterns. In deep learning models, weight matrices W are often high-dimensional. For instance, in fully connected layers or convolutional layers, the dimensions of the weight matrices can be very large. Given a weight matrix $W \in \mathbb{R}^{d \times k}$, LoRA approximates the weight update ΔW as a product of two low-rank matrices:

$$\Delta W = AB^T \tag{4.1}$$

where $A \in \mathbb{R}^{d \times r}$ and $B \in \mathbb{R}^{k \times r}$, and r is much smaller than both d and k. The updated weight matrix W' is then given by:

$$W' = W + \Delta W = W + AB^T \tag{4.2}$$

The primary objective of LoRA is to minimize the loss function L on the training data, where the model parameters are updated using the low-rank adaptation:

$$\min_{A,B} L = MSE\left(W_{pretrained} + AB^T\right) \tag{4.3}$$

Here, $\theta_{pretrained}$ represents the pre-trained model parameters, and A and B are the low-rank matrices. Gradient descent is used to update the parameters of the low-rank matrices A and B.

2.3 Performance Evaluation

To evaluate the daylight performance of floor plan images generated by diffusion models, the traditional method involves importing images into professional modeling software, establishing geometrical and physical models, applying materials, and running simulations. Current image-to-image prediction methods offer a solution to this problem.

Therefore, this study employs the pix2pixHD model, known for its strong performance in image-to-image translation, to efficiently evaluate the results generated by the diffusion model.

Dataset for Training Pix2pixHD

Daylight simulation was conducted using Ladybug and Honeybee on the Grasshopper platform in Rhino to generate image pairs for training the pix2pixHD model. We used the EPW weather file for Shanghai. The building geometry features window-to-wall ratios of 0.2 for the north-facing façade and 0.3 for the south-facing façade. The simulation parameters conform to the *Code for Thermal Design of Civil Buildings* (GB50176-2016) [22] and adhere to standard construction dimensions. Material properties were selected to reflect commonly used construction materials, as detailed in Tables 4.2 and 4.3.

Table 4.2 Simulation parameters

Parameters	Value
Window to wall ratio (north)	0.2
Window to wall ratio (south)	0.3
Windows height	1.8 m
Sill level height	0.9 m
Story height	3.3 m

Table 4.3 Material property in daylight simulation model

Opaque material	Reflectance
Ceiling	0.6
Wall	0.5
Floor	0.3
Glazing material	Transmissivity
Windows glaze	0.6

Spatial Daylight Autonomy (sDA) measures the percentage of a space that receives adequate daylight throughout the year. According to the Illuminating Engineering Society (IES), sDA300/50% indicates the proportion of points in an area that achieve a horizontal illuminance of at least 300 lx for 50% of occupied hours (8 a.m. to 6 p.m.) over a typical meteorological year. The sDA equation is as follows:

$$sDA_{x/y\%} = \frac{\sum_i^n P(i)}{\sum_i^n p_i} \in [0, 1]$$

$$\text{where } P(i) = \begin{cases} 1\,;\ if\ DA \geq DA_{limit} \\ 0\,;\ if\ DA \leq DA_{limit} \end{cases} \tag{4.4}$$

The sDA standard, as defined by IES Lighting Measurements (LM) 83–12 [23], uses the parameters "sDA 300, 50%" measured from 8 AM to 6 PM over the year. This metric evaluates the daylight performance of a space by categorizing it into three classes based on the percentage of the area meeting the sDA criterion, the details are listed in Table 4.4.

Table 4.4 Standard of spatial daylight autonomy

Level of recommendation	Minimum target illuminance/lx	Fraction of daylight hours/%	Fraction of spaces for target level/%
Preferred	300	50%	≥75%
Nominally accepted			55% ~ 75%
Low			≤55%

Model Training of Pix2pixHD

With the training dataset, we trained a pix2pixHD model to quick predict the daylight performance, which consists of two components: a coarse-to-fine generator and multi-scale discriminators. The generator of pix2pixHD is called coarse-to-fine generator, which can be decomposed into sub-networks, a global generator network G_1 and a local enhancer G_2. The global generator network G_1 is responsible for generating a coarse, low-resolution version of images. This network captures the overall structure and global features of the image but does not focus on fine details. The output from G_1 serves as the foundational layer upon which further refinements are made. To evaluate the generated images, pix2pixHD employs multi-scale discriminators at different scales. Each discriminator is responsible for assessing the realism of the image at a specific resolution, ensuring that both local details and the global structure are realistic.

The authors of Pix2PixHD improved upon the traditional conditional GAN loss $\mathcal{L}_{GAN}$ by incorporating the Feature matching loss $\mathcal{L}_{FM}$ and perceptual loss $\mathcal{L}_{VGG}$. The complete loss function consists of the following equations:

$$\mathcal{L}_{total}(G, D_k) = \sum_k \mathcal{L}_{GAN}(G, D_k) + \lambda_{FM} \sum_k \mathcal{L}_{FM}(G, D_k) + \lambda_{VGG}\mathcal{L}_{VGG}(y, G) \tag{4.5}$$

In this study, a total of 300 pairs of images were used for training, of which 80% were used for the training set and 20% for the test set. The pix2pixHD model was trained using the Adam optimizer with a learning rate of 0.0002, due to the time required per epoch, which spans several minutes, the maximum number of epochs was set to 200 to balance performance and computational cost. Both networks were updated at every step using loss functions, which guided the model to generate realistic data through iterative optimization. The loss weights were set to $\lambda_{FM} = \lambda_{VGG} = 10$.

Model Evaluation

To quantitatively determine the optimal epoch, we calculated the Fréchet Inception Distance (FID) scores for each epoch. The Fréchet Inception Distance (FID) is a metric used to assess the quality of generated images by comparing their statistical properties to those of real images. FID calculates the distance between the feature distributions of the generated images and real images, as extracted by an Inception v3 network. Lower FID scores indicate that the generated images are more similar to the real images.

After selecting the optimal model, we evaluate its performance using three metrics: Mean Squared Error (MSE), Peak Signal-to-Noise Ratio (PSNR), and Structural Similarity Index (SSIM). These metrics provide a comprehensive evaluation by measuring the pixel-level accuracy, perceptual quality, and structural fidelity of the generated images.

3 Results

3.1 Training Results of LoRA

Figure 4.3 illustrates the loss curve during LoRA training, spanning 48,000 steps across 10 epochs. A consistent downward trend in the loss indicates effective learning and adaptation by the model. The curve shows a steep decline in the initial epochs, gradually tapering off in later epochs, suggesting convergence. Minor fluctuations in the loss values, due to inherent variability in the training data and the stochastic nature of the AdamW8bit optimizer, remained within acceptable limits, demonstrating the stability and robustness of the training process. Notably, the loss reached its lowest point at approximately the 40,000th step (step 39,731). Given that each epoch consists of 4,800 steps, we selected the 8-th epoch for generating the building floor plan.

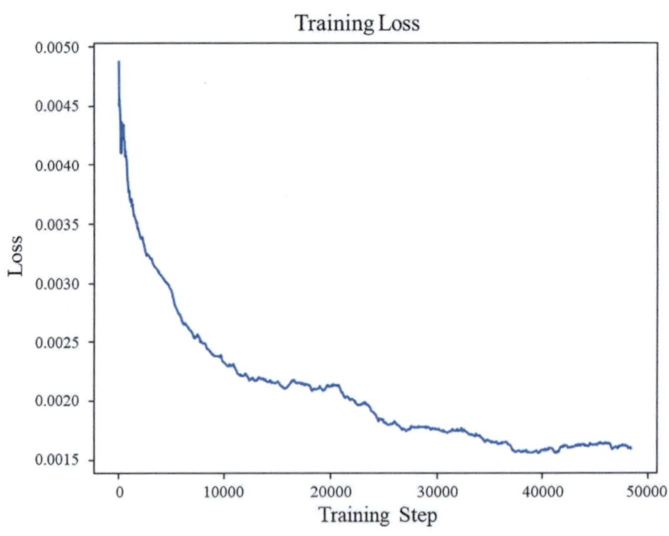

Fig. 4.3 Traning loss of LoRA

3.2 Training Results of Pix2pixHD

Figure 4.4 displays pairs of images at various training epochs for the model. We compared the synthesized images from epochs 10, 30, 50, 100, 150, and 200 with their corresponding ground truth images. Starting from epoch 30, the differences between the synthesized and ground truth images become difficult to distinguish through visual inspection.

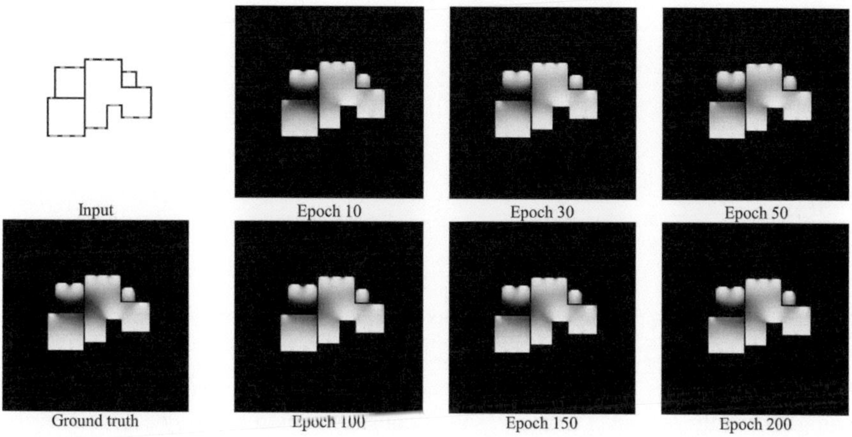

Fig. 4.4 Daylight performance prediction using pix2pixHD under different epochs

Shown in Fig. 4.5, we found that the model reached its lowest FID score of 12.85 at epoch 170. This indicates the best performance in terms of image quality and similarity to the ground truth. Consequently, the model from epoch 170(170_net_G) was selected for subsequent predictions, ensuring the highest fidelity in generated results.

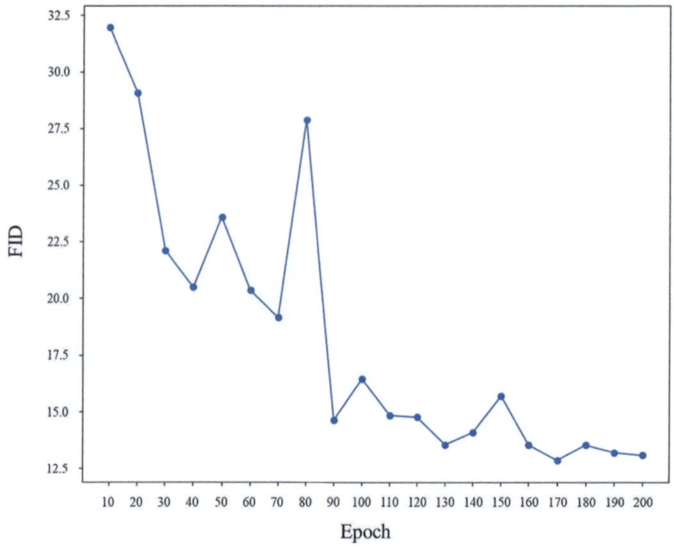

Fig. 4.5 FID score of pix2pixHD model at different epochs

Table 4.5 presents the performance evaluation of the selected model on both the training set and the test set. The model shows consistently strong results, with an average SSIM of 0.98 for both the training and test sets, approaching the ideal value of 1, which indicates a high level of structural similarity with the ground truth. The MSE scores are low, averaging 4.9 on the training set and 4.2 on the test set, reflecting minimal prediction error. Correspondingly, the PSNR scores, derived from MSE, are around 40 dB on the test set, signifying high-quality image generation.

Table 4.5 Performance metrics (MSE, PSNR, SSIM) on training and test sets

Dataset	Average of MSE	Average of PSNR	Average of SSIM
Training set	4.9	41.8	0.98
Test set	4.2	42.6	0.98

As shown in Fig. 4.6, to verify the accuracy of the model, we compare the predicted values generated by pix2pixHD with the sDA simulation results, and find that the predicted values deviate from the actual values within 5%, with a standard deviation of 2.08, which proves that our trained model is able to predict the daylight performance of planar maps well.

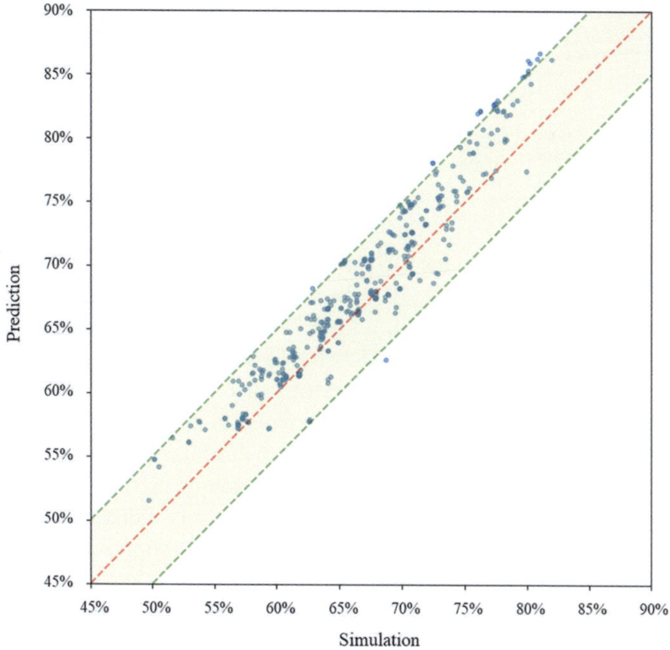

Fig. 4.6 Comparison of sDA of predicted values with ground truth

3.3 Floor Plans Generation and Evaluation

To evaluate the generation model's effectiveness, we produced 1,800 layouts, which were filtered using an OpenCV script. As shown in Table. 4.6, these layouts, each at a resolution of 1024 × 1024 pixels, were created by the diffusion model. After filtering, 89% closely matched the training set features and met key design constraints, while the remaining layouts had issues like misplaced windows, duplicated spaces, or incomplete partitions. Although post-processing could resolve these errors, this study primarily focuses on the diffusion model's ability to generate plausible layouts for early-stage design. The high success rate underscores the model's capability to capture essential architectural features, making it a valuable tool for rapid preliminary design in building development.

Table 4.6 Samples of generated results

Good case (89%)			Bad cases (11%)		
					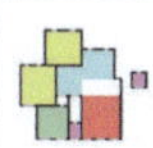

(*continued*)

Table 4.6 (*continued*)

Good case (89%)			Bad cases (11%)		

The floor plans generated by the diffusion model were assessed for Spatial Daylight Autonomy (sDA) using the pix2pixHD model. Most plans fell within the nominally accepted range (55% < sDA ≤ 75%), with 2,787 cases (65%) meeting this criterion. Conversely, 1,087 plans (25.3%) showed low daylight performance (sDA ≤ 55%), while only 415 plans (9.7%) achieved preferred performance (sDA > 75%). This distribution highlights the significant impact of floor plan layout on daylight performance, emphasizing the potential of automated generation tools to optimize designs. Figure 4.7 shows that as floor area increases, the percentage of plans achieving preferred sDA performance decreases, largely due to the complexities of larger layouts. For architects, automated tools can efficiently generate diverse design options, facilitating refinements based on daylighting outcomes.

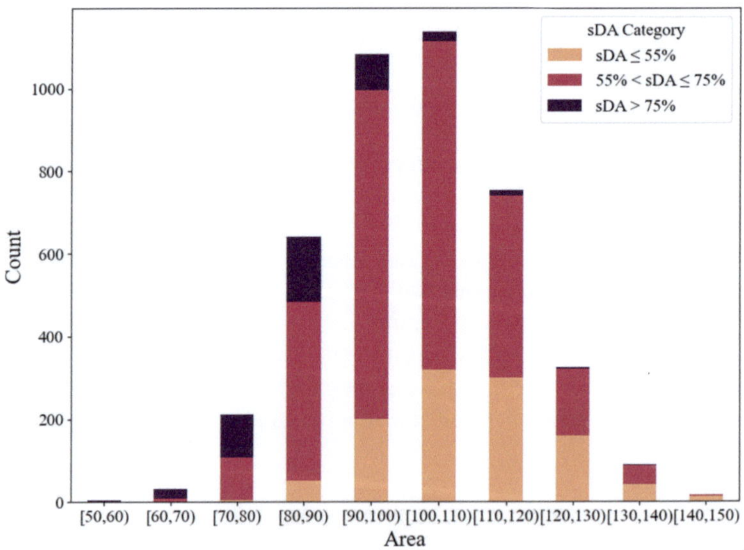

Fig. 4.7 sDA distribution of generated result

To evaluate the impact of spatial layout on daylight performance, representative floor plans were selected from four size ranges: [80,90), [90,100), [100,110), and [110,120). "Preferred" plans maximize daylight with a horizontal layout featuring north–south windows, while "nominally accepted" plans are more compact, sometimes resulting in inadequate lighting. "Low-performance" plans, with vertically stacked rooms, suffer from poor daylight due to limited east–west windows. Although higher spatial Daylight

Autonomy (sDA) is desirable, achieving over 75% sDA isn't always necessary. Wider, shallower layouts optimize natural light but can increase wind loads and reduce insulation, whereas narrower designs may limit lighting and ventilation. Most designs strike a moderate balance, though room layout variations can significantly affect performance. This basic analysis highlights the need for nuanced approaches that consider specific room requirements, suggesting future research should evaluate daylight performance based on occupants' daily routines (Table 4.7).

Table 4.7 Representative samples of generated floor plans and their corresponding daylight performance

Area	80–90 m^2	90–100 m^2	100–110 m^2	110–120 m^2
Preferred (sDA $\geq$ 75%)	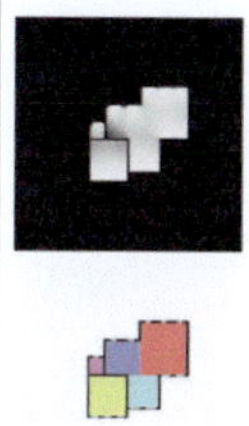 sDA=83%	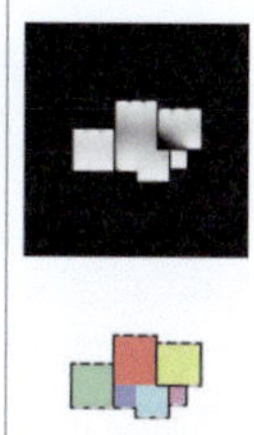 sDA=82%	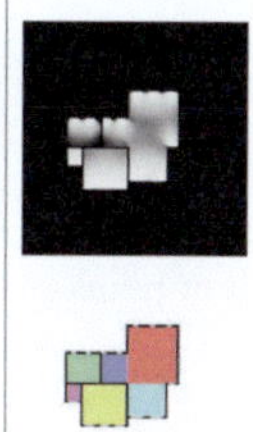 sDA=80%	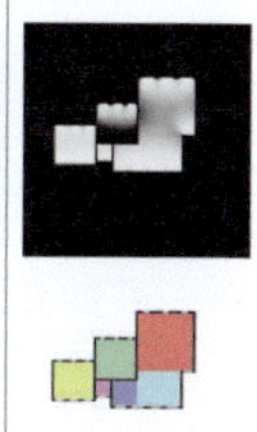 sDA=76%
Nominally accepted (75% $\geq$ sDA $\geq$ 55%)	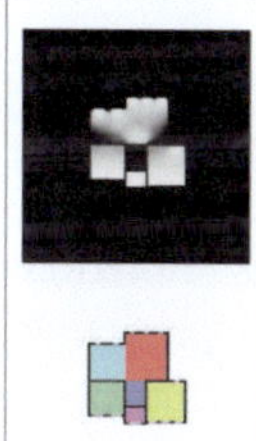 sDA=66%	 sDA=63%	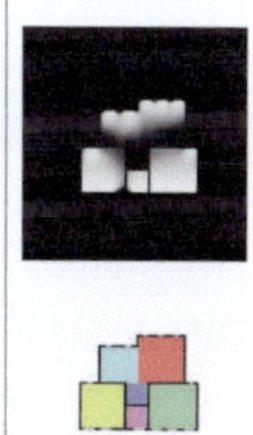 sDA=60%	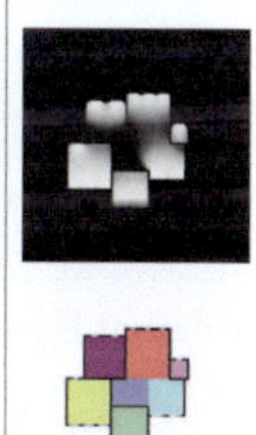 sDA=58%
Low (sDA $\leq$ 55%)	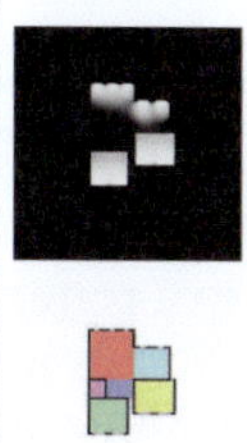 sDA=51%	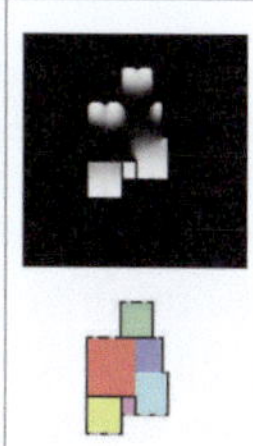 sDA=44%	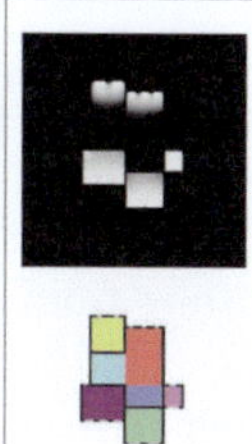 sDA=44%	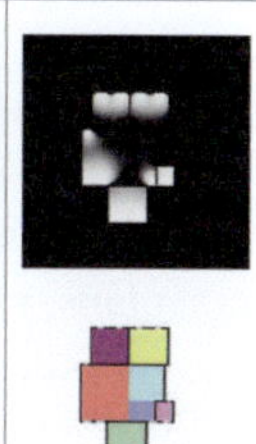 sDA=40%

4 Discussion

Architectural design is complex, requiring a balance of multiple stakeholders' needs, and cannot be fully replaced by automated layout generation tools. However, these tools excel in providing feedback, design guidance, and optimization during the initial stages by generating various design options and identifying promising solutions. Our method generates relatively simple floor plans that offer essential layout information, facilitating relevant guidance in the early-stage design. With just a basic text prompt, it simplifies the design process and prioritizes early-stage guidance focused on performance and environmental factors. Notably, our approach demonstrates significant efficiency, as the pix2pixHD model evaluates performance 267 times faster than traditional daylight simulation methods (Table 4.8).

Table 4.8 Time costs of floorplan generation and performance evaluation

Task	Test	Training time	Calculation time per case
Generation of floor plans	Rhino + Grasshopper	N/A	15 ~ 30 s
	Diffusion model + LoRA (1024*1024)	353 min	12 s
Performance evaluation	Daylight simulation (grid size: 0.1)	N/A	6 min 14 s (374 s)
	Pix2pixHD	423 min	1.4 s

Despite its advantages, our method has limitations. Improved controllability and reliability in floor plan generation are needed, particularly regarding room types, sizes, and finer architectural details like window placement. Additionally, the current performance prediction is limited to specific and simple floor plan styles, affecting the generalizability of the methodology.

5 Conclusion and Future Work

This study integrates diffusion models and GANs to optimize AI-generated architectural floor plans, shifting from a form-driven to a performance-driven approach that better meets architects' practical needs. By significantly reducing the time for iterative design modifications and performance simulations, the method improves efficiency in the early stages of design.

We implemented performance evaluation using the pix2pixHD model, allowing rapid predictions of daylight performance. This integration combines AI's creative generation with performance evaluation, providing architects with informed design solutions that meet both creative and performance criteria. Our learning-based approach offers a remarkable speed advantage over traditional rule-based methods, reducing performance simulation time from 6 min and 14 s to just 1.4 s—an impressive 267-fold improvement.

However, our model faces limitations regarding the controllability of generative outputs and the generalizability of predictions. Future work will focus on incorporating real-time interaction tools and expanding the variety of floor plan types to enhance effectiveness and applicability.

References

1. Weber RE, Mueller C, Reinhart C (2022) Automated floorplan generation in architectural design: a review of methods and applications. Autom Constr 140:104385
2. Rahbar M, Mahdavinejad M, Markazi AHD, Bemanian M (2022) Architectural layout design through deep learning and agent-based modeling: a hybrid approach. J Build Eng 47:103822
3. Stiny G (1980) Introduction to shape and shape grammars. Environ Plann B Plann Des 7(3):343–351
4. Arvin SA, House DH (1999) Making designs come alive: using physically based modeling techniques in space layout planning. In: Augenbroe G, Eastman C (eds) Computers in building. Boston, MA: Springer US, pp 245–62. https://doi.org/10.1007/978-1-4615-5047-1_16
5. Guo Z, Li B (2017) Evolutionary approach for spatial architecture layout design enhanced by an agent-based topology finding system. Front Arch Res 6(1):53–62
6. Bao F, Yan DM, Mitra NJ, Wonka P (2013) Generating and exploring good building layouts. ACM Trans Graph 32(4):1–10
7. Merrell P, Schkufza E, Koltun V (2010) Computer-generated residential building layouts. ACM Trans Graph 29(6):1–12
8. Goodfellow I, Pouget-Abadie J, Mirza M, Xu B, Warde-Farley D, Ozair S et al (2020) Generative adversarial networks. Commun ACM 63(11):139–144
9. Wu W, Fu XM, Tang R, Wang Y, Qi YH, Liu L (2019) Data-driven interior plan generation for residential buildings. ACM Trans Graph 38(6):1–12
10. Nauata N, Hosseini S, Chang KH, Chu H, Cheng CY, Furukawa Y (2021) House-GAN++: generative adversarial layout refinement network towards intelligent computational agent for professional architects. In: 2021 IEEE/CVF conference on computer vision and pattern recognition (CVPR), pp 13627–36. Nashville, TN, USA: IEEE. https://ieeexplore.ieee.org/document/9577959/
11. Yi H, Yi YK, Chan T (2014) Performance based architectural design optimization: automated 3D space layout using simulated annealing. In: Proceedings of the 2014 ASHRAE/IBPSA-USA building simulation conference, Atlanta, GA, USA, pp 10–2
12. Isola P, Zhu JY, Zhou T, Efros AA (2017) Image-to-image translation with conditional adversarial networks. In: Proceedings of the IEEE conference on computer vision and pattern recognition, pp 1125–34
13. Ho J, Jain A, Abbeel P (2020) Denoising diffusion probabilistic models. Adv Neural Inf Process Syst 33:6840–6851
14. Rombach R, Blattmann A, Lorenz D, Esser P, Ommer B (2022) High-resolution image synthesis with latent diffusion models. In: Proceedings of the IEEE/CVF conference on computer vision and pattern recognition, pp 10684–95
15. Shabani MA, Hosseini S, Furukawa Y (2023) HouseDiffusion: vector floorplan generation via a diffusion model with discrete and continuous denoising. In: 2023 IEEE/CVF conference on computer vision and pattern recognition (CVPR), pp 5466–75. Vancouver, BC, Canada: IEEE. https://ieeexplore.ieee.org/document/10204517/
16. Zeng P, Gao W, Yin J, Xu P, Lu S (2024) Residential floor plans: multi-conditional automatic generation using diffusion models. Autom Constr 162:105374

17. Li X, Yuan Y, Liu G, Han Z, Stouffs R (2024) A predictive model for daylight performance based on multimodal generative adversarial networks at the early design stage. Energy Build 305:113876
18. He Q, Li Z, Gao W, Chen H, Wu X, Cheng X et al (2021) Predictive models for daylight performance of general floorplans based on CNN and GAN: a proof-of-concept study. Build Environ 206:108346
19. Mokhtar S, Sojka A, Davila CC (2020) Conditional generative adversarial networks for pedestrian wind flow approximation. In: Proceedings of the 11th annual symposium on simulation for architecture and urban design, pp 1–8
20. Wang TC, Liu MY, Zhu JY, Tao A, Kautz J, Catanzaro B (2018) High-resolution image synthesis and semantic manipulation with conditional GANs. In: 2018 IEEE/CVF conference on computer vision and pattern recognition. Salt Lake City, UT, USA: IEEE, pp 8798–807. https://ieeexplore.ieee.org/document/8579015/
21. Hu EJ, Shen Y, Wallis P, Allen-Zhu Z, Li Y, Wang S et al (2021) LoRA: low-rank adaptation of large language models. http://arxiv.org/abs/2106.09685
22. GB50176 C (2016) Code for thermal design of civil building. Ministry of Housing and Urban-Rural Development of the People's Republic of China: Beijing, China
23. Heschong L, Wymelenberg VD, Keven, Andersen M, Digert N, Fernandes L et al (2012) Approved method: IES spatial daylight autonomy (sDA) and annual sunlight exposure (ASE). https://api.semanticscholar.org/CorpusID:114427665

A Machine Learning Method for Predicting the Spatial Combination of Community-Embedded Retirement Buildings

Zhiqi Wang, Yangpeng Xin(✉), Ying Zhou(✉), and Mingyu Cheng

Southeast University, Nanjing, China
{213210126,zhouying}@seu.edu.cn

Abstract. Against the background of actively responding to population aging, the demand for community-embedded retirement building construction is increasing daily. Given the density and complexity of urban built-up areas, renovating existing community buildings in the context of inventory renewal has opened up a practicable development path for retirement building construction. However, the contradiction between the spatial structure of existing buildings and the functional requirements of retirement buildings has become a difficult problem. The environment of urban communities in China is diverse and complex, and the traditional program that relies solely on architects or community managers to carry out the renewal design has the shortcomings of low efficiency and low utilization rate. In this paper, we propose a spatial combination prediction method for retirement buildings based on a Graph Convolutional Neural network (GCN), which realizes the prediction of spatial combinations of retirement buildings by using the GCN node classification task model and the generative design method under the constraints of the existing site environment. This study also takes the building room characteristics and room connectivity into consideration, which can reflect the building usage more realistically in model training and generative design optimization, and this method has significant advantages in both time efficiency and space utilization.

Keywords: Retirement building · Graph convolutional neural network · Generate design · Space combination

1 Introduction

China's aging population is rapidly increasing, making it a national priority to address this issue and provide basic elderly care services for all. This demographic shift has led to a sharp rise in demand for elderly care services, exacerbating the supply–demand gap. With the miniaturization of family size and rapid economic and social development, home-based elderly care is becoming less feasible. Consequently, community-embedded elderly care is emerging as a key direction for China's modern elderly care model [1], driving up the demand for such facilities. In dense urban areas, renovating existing community buildings offers a practical solution for developing retirement facilities.

J. Fu et al. (eds.), *AI for Architecture*, Computational Design and Robotic Fabrication,
https://doi.org/10.1007/978-981-95-0974-4_5

However, the contradiction between the spatial combination of existing buildings and the functional needs of retirement buildings has become a significant challenge [2]. In China's diverse and complex urban communities, relying solely on architects or community managers for renewal design is inefficient and results in low utilization. While some sustainable community and building renovation methods were discussed in studies—such as community renewal strategies based on "active aging" theory [3], decision support systems for residential building renovation [4], and the "middle zone" optimization design method for interior space of retirement buildings [5], there remains a lack of effective methods to integrate retirement building spaces into the existing community spatial combination.

This paper proposes a spatial combination prediction method based on graph convolutional neural network (GCN) for retirement buildings, using the GCN node classification task model and generative design to predict spatial combinations under existing site constraints. GCN, an emerging deep learning model for graph-structured data, excels in analyzing and predicting complex graph-structured nodes but is less applied in architecture. In this model, the building spatial structure resembles a functional bubble graph, translatable into graph-structured data with node characteristics. Generative design, a key development in building digital technology, remains vital due to emerging optimization algorithms. This study considers existing site limitations, using generative design algorithms and graph structure optimization to predict feasible retirement building space combinations. It also accounts for building room characteristics and connectivity, enhancing the model's realism in usage prediction and generative design optimization, with significant advantages in time efficiency and space utilization.

The paper introduces the study's background and methodology, followed by a literature review on community-embedded retirement buildings, community environment renovation, GCNs, and automatic floor plan generation design. It then details the technical methodology and findings, concluding with a discussion.

2 Literature Review

2.1 Community-Embedded Retirement Building Construction

With the aggravation of population aging and the diversified needs of the elderly, the construction of community-embedded retirement buildings is an important initiative to respond to the challenges of the aging society, meet the needs of the elderly for senior care services, and promote the innovation of senior care service models. Several studies have used case studies, field research, statistics, and modeling analysis to propose strategies related to the construction of community-embedded retirement buildings, such as Hu Huiqin et al. pointed out that converting residences into retirement buildings is an effective construction pathway through a field study of the background Chaoyang District's aging communities [6], and Zhang et al. under the concept of CCRC, through a case study and statistical methods proposed a spatial construction framework of facility units, flexible organization and integrated implantation, and proposed a spatial configuration strategy of facility types, functional organization and allocation methods based on the research results [7]. Some scholars have also studied the demand and satisfaction of community-based home care services through the Kano model analysis and AHP-fuzzy comprehensive evaluation method, and based on embeddedness theory, proposed tangible and intangible embeddedness modules based on the embedded theory [8]. The embedding method and scale of retirement buildings in urban established settlements need to be determined by community environment data [9].

As architects and designers, urban small-scale embedded retirement buildings should meet the principles of operational logic and integration into the community, and establish the value standard of "people-oriented" [10], and there is a correlation between the spatial combination of the retirement building and the living behaviors of the occupants [11], so the research on the spatial combination of the retirement building should be emphasized. Focus on the study of the spatial structure of retirement buildings, which inspires this study. In addition, in the context of the rapid development of intelligent technology, intelligent design strategies have been gradually applied to retirement building design [12]. In this study, the new deep learning technology in the field of artificial intelligence guides the direction of the research problem, takes the spatial combination of retirement buildings as an important object of research, adheres to the value orientation of "human-centeredness", and promotes the theoretical progress of intelligent design strategies in the field of retirement building research.

2.2 Renovation of Existing Environments in the Community

In the context of the era of inventory renewal, along with the aging of the population and the built environment of the community, the community environment and existing building renovation become more and more necessary. Professor Cheng Xiaoqing led a team to carry out the practice of aging-adapted renovation in the Beijing Dazhalan community in 2015 and put forward renovation suggestions and methods in combination with practical experience [13], and some scholars summarized that the replacement method, alteration method, and addition method are of guiding significance in the use of existing resources to adapt to the new functional needs [2]. At the community level, the Kano model helped to explore the supply and demand of community elderly services and the design needs of community public spaces [14]. Han Linfei et al. proposed a community renewal strategy based on the theory of "active aging", which includes the reconstruction of spatial systems and the establishment of an integrated elderly management system [3].

Traditional research methods, such as those based on the Kano model, undoubtedly have great potential, but the complexity of the work process and the multiple objectives of the research make it difficult to relate to the spatial design level. The research methodology in this paper focuses on architectural spatial combinations, making the theoretical approach operational and practicable.

2.3 Graph Convolutional Neural Networks (GCNs)

Graph data consists of nodes and edges, which can contain node features and edge weights, and is commonly found in fields such as social networks, recommender systems, biological information, etc. Graph convolutional neural networks are a class of deep learning models for dealing with graph data, which have received wide attention in recent years and have been widely used in computer vision, biochemistry, natural language processing, and recommender systems, etc. [15]. As a powerful tool for analyzing graph data, graph convolutional neural network has obvious advantages in node classification, graph classification, edge prediction, etc. In the field of architecture, some scholars provided an efficient solution for beam layout design of frame structures by using graph convolutional neural network based on the topological characteristics of beam-column connections of frame structures, and the model prediction results are not much different from the design results of structural designers [16]. The adjacencies of different rooms in a building were difficult to summarize in a nutshell, and a method proposed to extract the spatial topology from BIM and represent it as a graph, and in turn, utilized a subgraph matching algorithm to achieve fast querying in the early stage of design [17]. In this paper, we refer to and simplify such topology graphs and utilize GCN to perform convolution operations and feature extraction on them.

2.4 Automatic Floor Plan Generation Design

With the improvement of computing power and the development of artificial intelligence technology, architectural generative design gradually plays an important role in the field of digital architecture, especially in the automatic generation of architectural floor plans to highlight the charm. Weber Ramon Elias et al. from MIT combed through the mainstream computerized architectural automatic layout methods and attributed the architectural automatic floor plan generation design methods to three categories: bottom-up, top-down, and referential [18]. For a long time, GAN algorithms were generalized for floor plan generation research. Cui et al. explored the learning effect of GAN generative models with different architectures on generating class floor plans [19], and Liu et al. applied GAN algorithms to functional floor plan generation of a specific type of building using knowledge graphs [20], the retirement building floor plan is functionally diverse and complex, the GAN algorithm has not been able to be applied to the design of automatic floor plan generation for retirement buildings.

Hua proposed a mathematical method [21] to understand the building floor plan layout as two parts: independent rooms and relationships between rooms, focusing on the irregular shape assessment of each room and the neighboring relationship between each pair of rooms, which provides the idea revelation and theoretical support for the progress of the generative design phase of this study.

3 Methodology

In this study, we propose a spatial combination prediction method for community-embedded retirement buildings using deep learning (GCN) and generative design, as shown in Fig. 5.1. Using real community retirement building plans as the database and research object, we first depict the dataset with Rhino® and observe the spatial combination characteristics. Grasshopper is used to get features from the dataset, and the Random Forest Algorithm classifies the standardized node features. We then build a graph convolutional neural network model for node classification. Finally, using the trained GCN model combined with a genetic algorithm, we predict the spatial combinations of retirement buildings within existing building plans.

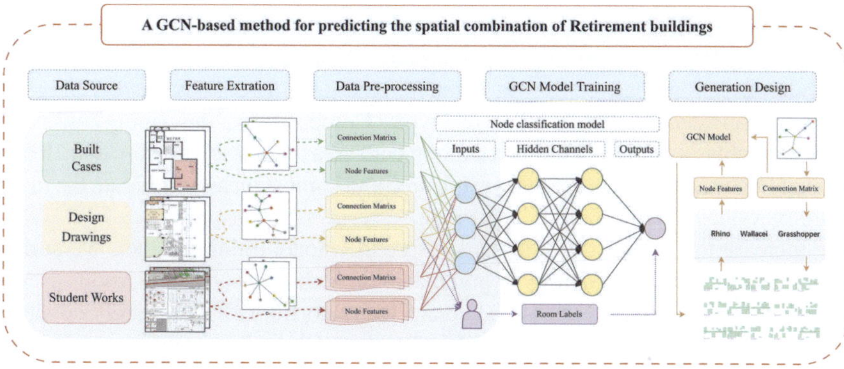

Fig. 5.1 The technical framework of this study, self-drawing

3.1 Data Acquisition and Pre-processing

To enhance the generalization ability of the graph convolutional neural network model, ensuring dataset diversity is essential. We collected 39 building plans from various sources, including community-embedded retirement buildings, community home care service facilities, community daycare centers, and student works in retirement communities. Through Rhino and Grasshopper, the area, perimeter, and coordinates of rooms were obtained, along with data on room connectivity, Table 5.1 shows some of the data we acquired.

Table 5.1 Sample dataset, self-drawing

Index	Source	Area	Perimeter	X	Y	Connection	Type
1	18(Built Case)	20.67	18.98	−12.48	−5.45	3,7	5(Bedroom)
15	17(Built Case)	6.79	16.30	12.40	1.76	19	4(Service)
320	6(Student Work)	4.79	8.80	39.62	−17.20	324	4(Service)
543	2(Student Work)	48.79	28.18	1.32	−14.45	547	3(Activities)
938	27(Design Plan)	105.58	46.39	21.45	−2.98	934,945	1(Public)
1111	31(Design Plan)	29.58	23.07	10.78	−45.83	1119	3(Activities)
1339	22(Design Plan)	178.38	103.24	−2.74	−3.85	X	2(corridor)

In building plans, “doors” signify traversal between spaces. Rooms are represented as nodes with features, and connections or doors between rooms are edges, effectively described by an adjacency matrix. The datasets in this study, derived from existing retirement building plans, are free from missing or outlier points. Feature engineering explored correlations between room features, types, and connectivity, revealing high correlations between room area, shape, and type. "Compactness," describing how closely a shape resembles a circle, was selected along with area as characterization data. Robust normalization was used to preprocess room area and compactness to mitigate the impact of extreme values.

To increase the dataset diversity and ensure the model training effect. We perform data augmentation on the existing 39 graph data. In architectural design, room-to-room connections are specific, foyers can be connected to corridors but are rarely connected to bedrooms. Assuming that corridor A1 is connected to bedroom A2 in graph A, and that corridor B2 is connected to bedroom B2 in graph B, then we assume that it holds that A1 is connected to B2 in diagram C. Using this permutation method, we expand the original 39 graph data to 117, and participate in graph convolutional neural network training together.

3.2 Graph Convolutional Neural Network Model Building

Graph Convolutional Neural Networks (GCNs) excel in node classification by aggregating neighboring features. The GCN model construction starts with data preparation, obtaining the node feature matrix and edge index. The node feature matrix contains each node's feature vector, while the edge index describes node connectivity. The model comprises multiple graph convolution layers, updating node features based on neighboring node information.

In this method, each layer's node features incorporate information about both the node itself and its neighbors. We build a three-layer graph convolutional neural network (GCN) to capture the graph's complex relationships more effectively. Initially, a random forest algorithm evaluates the impact of features on node classification, calculates the weights for each feature, and multiplies these weights with the node features before feeding them into the convolutional layer.

Each graph convolutional layer's output features serve as the input for the next layer, with the final layer's output used for classification. During forward propagation, node features and edge indices are input into each graph convolution layer, followed by a ReLU activation function to generate updated node representations. The Adam optimizer adjusts each layer's parameters via backpropagation to minimize discrepancies between model predictions and true labels. The training process includes validation steps to monitor model performance, prevent overfitting, and adjust hyperparameters.

Node and edge indices are mapped and filtered to ensure only the correct training, validation, and test data are used. The ultimate goal is to achieve good classification performance on unseen test data through iterative training and validation. Key aspects of the model-building process include node feature updates, information aggregation in each graph convolution layer, parameter optimization, and performance evaluation during training, as illustrated in Fig. 5.2. These components together form an end-to-end GCN model.

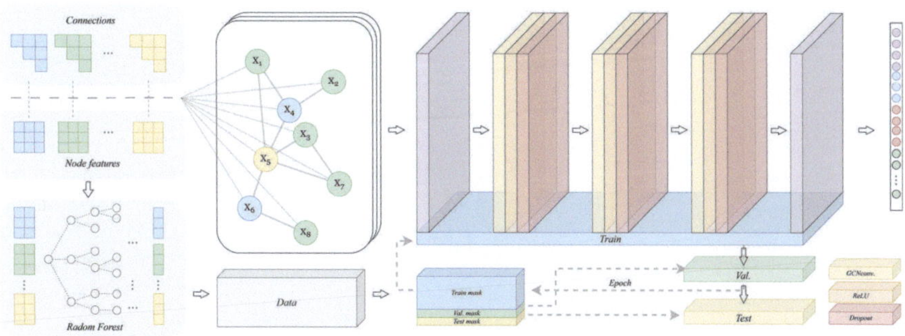

Fig. 5.2 GCN-based model architecture for node classification, self-drawing

3.3 Design of Spatial Combination Plan Generation for Retirement Building

Automatic floor plan generation design with room adjacencies as a condition [21] is a method that can be developed among retirement buildings. In this study, the floor plan layout of the building is divided into two parts: individual rooms and relationships between rooms. The focus is on evaluating the characteristics of each room and the adjacencies between pairs of rooms. According to the shape of the input image, the program generates various layouts that satisfy the user's geometric and topological requirements.

The user inputs the spatial layout of the existing building plan and the expected adjacencies, and the program calculates the topological cost by comparing the adjacencies between the desired adjacencies and the actual areas. This work is based on the Rhino platform and grasshopper programming and is implemented using a Genetic Algorithm through Wallacei, which evaluates the geometric feature cost of each room by the GCN model that has been trained to take into account the shape and size of the room. Finally, the floor plan layout solution that best meets the desired conditions is filtered. Figure 5.3 illustrates a programming scheme based on Rhino, Grasshopper, and Wallacei that sets a random seed for the selection of an existing graph.

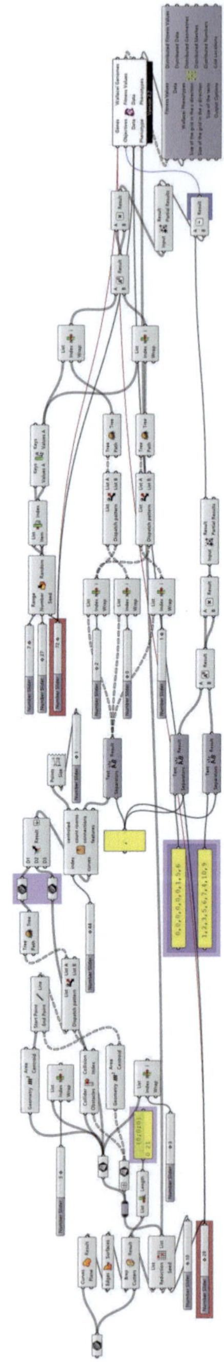

Fig. 5.3 A grasshopper-based implementation approach, self-drawing

4 Results and Discussions

This GCN model consists of three graph convolution layers, each of which updates the node representation by projecting node features into a new representation space and aggregating information from neighboring nodes. Each graph convolution layer is followed by a ReLU activation function to enhance the nonlinear modeling capability of the model. The model uses the Adam optimizer to optimize the model parameters with a learning rate of 0.0005, while L2 regularization (weight decay) is applied to control the model complexity.

After about 7 thousand epochs of training, the model has a final loss of 0.50 on the training set and 0.78 on the validation set, which indicates that the model can effectively learn the features of the data during training and performs well on the validation set, albeit with some fluctuations. Finally, on the unseen test data, the model has a test accuracy of 0.80. This result shows that our model maintains good classification performance even when dealing with new data, thus validating its generalization ability and usefulness.

We further analyze the training process of the model by plotting the image of the loss function with epoch. Figure 5.4 shows that the training loss and validation loss gradually decrease with the training process, and the validation loss also exhibits some fluctuations, but the overall trend remains consistent.

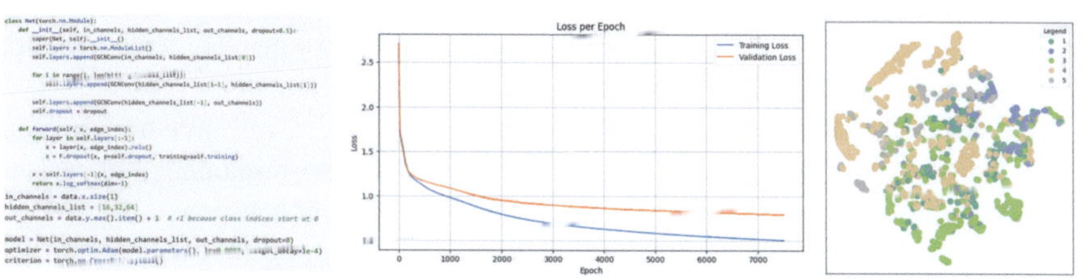

Fig. 5.4 Model definition, hyperparameter settings, loss function images, and training results, self-drawing

We select the first floor plan of a building to be remodeled in a certain community, whose original function is residential, and conduct a generative design experiment on it using the method in this study. First of all, the user wants to build a small community-based home care service center, which contains about 11 rooms, and the types of rooms and connectivity relationships are represented as a graph. We import this requirement and the existing building plan into the generative design program, which proposes several possible scenarios for the user, filters the expected scenarios by using adjacency relationships and the GCN model, and ultimately selects the one that best fits the user's requirement. We develop the choice and get the design result as shown in Fig. 5.5.

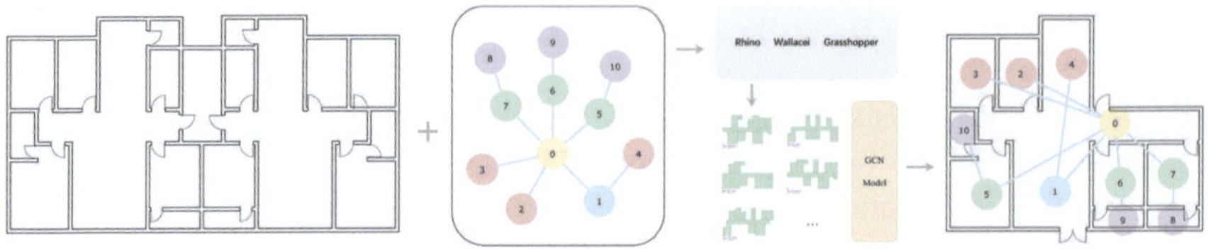

Fig. 5.5 An experiment in generative design, self-drawing

In the past, graph convolutional neural networks have been widely used in computer vision, biochemistry, natural language processing, recommendation systems, etc., while in the field of architectural design, this novel neural network technique has not been emphasized by researchers. This study incorporates the graph convolutional neural network approach into the exploration of architectural design, which is innovative at the theoretical level.

5 Conclusions

This paper proposes a spatial combination prediction method for community-embedded retirement buildings, using a mature retirement building floor plan layout scheme as the dataset and applying graph convolutional neural networks for deep learning. It then employs a generative design method to predict reasonable floor plan layouts considering the constraints of the established community environment.

This study has some limitations: (1) Regarding the dataset, resource and time constraints led to dataset selection and production based on the researcher's experience, causing certain biases and contingencies. Despite using data enhancement methods, the dataset's number and diversity are somewhat lacking, affecting model generalization and accuracy. (2) Regarding visualization, the proposed method is a pre-design spatial layout approach, but due to energy and time constraints, it has not yet been developed into an end-user application for designers or community managers, requiring further optimization and experimentation for practical application.

The expansion of the dataset and the design of the software visualization are of great significance for the future application of this study. By introducing more fine-grained retirement building data and field research, the characteristics and patterns of the spatial layout of community-embedded retirement buildings can be captured more precisely.

Future research will explore and innovate in the application of deep learning methods for architectural spatial assemblage, improve the design efficiency of community recreational facilities, and develop software visualization tools to contribute to the effective construction of community recreational facilities and the sustainable development of China's aging society.

Acknowledgements. This study was funded by the National Key Research and Development Program of China (2022YFF0607003), the National Innovation Training Program of Southeast University (202410286001Z), and the Key Innovation Training Program of Southeast University (202401012).

References

1. Du P, Ma Q (2024) Community-embedded elderly care in China: an analysis of the current situation and problems. Popul Dev 30(3):113–124
2. Li SJ, Jia YY, He WJ (2024) A study on the strategy of converting existing buildings into retirement buildings in urban communities under the background of inventory renewal. Architecture and Culture 2:157–159
3. Han L, Wang B (2023) Research on the community renewal of "Active Adaptation to the Elderly" in the unit compound. In: Proceedings of ECSF 2021, pp 1–14. Springer International Publishing
4. Serrano-Jiménez A, Barrios-Padura Á, Molina-Huelva M (2018) Sustainable building renovation for an aging population: decision support system through an integral assessment method of architectural interventions. Sustain Cities Soc 39:144–154
5. Wang H, Hou K, Kong Z, Guan X, Hu S, Lu M, Piao X, Qian Y (2022) "In-Between Area" design method: an optimization design method for indoor public spaces for elderly facilities evaluated by STAI, HRV, and EEG. Buildings 12(8):1274
6. Hu H, Xu Z, Che G (2018) A methodological exploration of converting residences into residential care facilities for the elderly in old communities—a case study of Chaoyang District, Beijing. J Asian Arch Build Eng 17(3):409–416
7. Zhang X, Liu C, Yang H (2022) Exploring the spatial configuration of community-embedded elderly care facilities under the CCRC concept: Building a continuum of care community in Tianjin. South Arch 5:73–82
8. Hong J, He Y, Long L (2023) A study on the demand and satisfaction of community home-based elderly services based on embeddedness theory: a case study of a community in Guilin city. In: People's city, planning for empowerment—proceedings of the annual conference on urban planning in China 2023 (19 Housing and Community Planning), vol 16
9. Xing Z (2023) Study on the embedding and scale of elderly service facilities in existing urban settlements—taking Beilin District of Xi'an as an example. In: People's city, planning empowerment—proceedings of the annual conference on urban planning in China 2022 (19 Housing and Community Planning), vol 14
10. Zhang S, Zhao Y (2017) Research on the design of embedded senior living facilities in small urban communities. J Arch 10:18–22
11. Li B, Li QL (2011) A comparative study of spatial structure and expansion of living behavior in senior living facilities. J Arch S1:153–159
12. Bu DQ, Zhang L, Jiang L (2024) Intelligent research on community home care service facilities. Intell Build Smart Cities 5:155–157
13. Cheng XQ, Zhang HX, Yin SZ (2018) Community practice of aging-adapted renovation of existing buildings: insights from the construction of a community retirement service station in Dazhalan, Beijing. J Arch 8:62–67
14. Lyu L, Tang H, Wang B (2024) Research on the renewal design of community public space under the smart elderly care model. In: Proceedings of the 2023 international conference on green building, civil engineering and smart city, pp 875–891. Springer Nature
15. Liu JQ, Tu WX, Zhu E (2023) A review of graphical convolutional neural networks. Comput Eng Sci 45(8):1472–1481
16. Zhao P, Liao W, Huang Y, Lu X (2023) Intelligent beam layout design for frame structure based on graph neural networks. J Build Eng 63:105499
17. Langenhan C, Weber M, Liwicki M, Petzold F, Dengel A (2013) Graph-based retrieval of building information models for supporting the early design stages. Adv Eng Inform 27(4):413–426

18. Weber RE, Mueller C, Reinhart C (2022) Automated floorplan generation in architectural design: a review of methods and applications. Autom Constr 140:104385. https://doi.org/10.1016/j.autcon.2022.104385
19. Cui Z, Guo Y, Li H (2023) A pix2pix-based method for generating and evaluating class plans. Resid Sci Technol 43(9):27–35
20. Liu J, Qiu Z, Wang L, Liu P, Cheng G, Chen Y (2024) Intelligent floor plan design of modular high-rise residential building based on graph-constrained generative adversarial networks. Autom Constr 159:105264. https://doi.org/10.1016/j.autcon.2023.105264
21. Hua H (2016) Irregular architectural layout synthesis with graphical inputs. Autom Constr 72:388–396

A Machine Vision Based Computational Approach to Micro-Behavior in Public Space

Te Li, Yanzi Jin, Jiaxin Yao, and Chao Yan(✉)

Tongji University, Shanghai, China
yanchao@tongji.edu.cn

Abstract. Social behavior is a core subject in public space research, yet its quantitative representation remains a significant challenge. With the advancement of machine vision technology, automated tracking and quantitative collection of spatial behaviors have become achievable. However, computational analysis and quantitative description methods for micro-behavioral data are still lacking. This study introduces a machine vision-based behavior computation method using pose estimation algorithms, aiming to develop a workflow for the quantitative analysis of micro-behaviors in public space scenarios. The feasibility of video-based micro-behavior computation tools is demonstrated. Focusing on micro-gait analysis as an example, this research reconstructs behavioral sequences and conducts empirical experiments to determine the optimal average shooting angles and distances, thereby improving the accuracy of data collection and behavior analysis. Additionally, a Grasshopper module is developed for behavioral sequence computation, enabling the visualization of gait metrics for micro-behaviors based on 17 joint data points. The accuracy of the machine vision-based behavior reconstruction method is further validated through motion capture experiments. Finally, using a social distancing experiment as a case study, this study provides preliminary evidence that the proposed method enhances the precision of micro-behavior analysis in complex public spaces. It demonstrates the potential to effectively represent public space indicators, such as social distancing.

Keywords: Behavioral analysis · Machine vision · Human pose estimation · Behavioral calculations · Micro-behavior

1 Introduction

The study of social behavior is a crucial area within behavioral performance research, highlighting the interactions among spatial users in specific built environments and their impact on daily experiences. Social behavior research commonly falls into two broad categories based on the goals of spatial evaluation and optimization. One category focuses on enhancing spatial quality and fostering vibrancy, emphasizing behaviors such as social interactions [1], recreational activities [2, 3], and consumer behaviors [4]. The other category prioritizes improving spatial efficiency and managing risks, with a focus on behaviors like pedestrian flow coordination [5], and close-contact interactions [6].

J. Fu et al. (eds.), *AI for Architecture*, Computational Design and Robotic Fabrication,
https://doi.org/10.1007/978-981-95-0974-4_6

Current research methods have significantly improved the efficiency of data collection compared to traditional approaches such as manual observation and questionnaire-based surveys. By utilizing technologies like WiFi, mobile signaling, and GPS for tracking the spatiotemporal trajectories of individuals, these methods have expanded the scale of studies on human movement patterns within spaces while minimizing interference with study subjects during observation [7, 8]. However, these approaches are limited to capturing location information, enhancing the breadth and efficiency of data acquisition but failing to provide deeper insights into micro-level behavioral information. As a result, fine-grained behavioral studies still face challenges in quantitative representation. With advances in observation and computation technologies, particularly those driven by computer vision, automated micro-behavior computation research has now become feasible.

2 Machine Vision-Based Research on Social Behavior

Computer vision methods enable the automated processing of video recordings. Algorithms such as object detection and object tracking can identify pedestrians and perform counting tasks [9]. Similar to the previously mentioned location-based analysis methods, approaches based on object detection and tracking allow human behavioral data to be recorded and analyzed on a large scale, in a contactless and automated manner. This eliminates experimental errors caused by the Hawthorne Effect and significantly improves computational efficiency. Moreover, video-based methods provide a more visualized computation process. They offer fast response times, enabling real-time monitoring of pedestrian flow data [10]. This method serves as a foundation for further acquisition of micro-behavioral data.

Another line of research employs depth cameras (RGB-D) to collect behavioral data. Relevant studies utilize wearable depth cameras mounted on participants to capture individual activities and interpersonal interaction information [11, 12]. This method enables the analysis of diverse behavioral information and the acquisition of spatial depth data. By integrating commonly used human pose estimation algorithms from depth camera development tools, it becomes possible to identify the specific behaviors of interacting individuals. With the advancement of deep learning technologies, human pose estimation algorithms based on standard monocular cameras have significantly improved in both speed and accuracy. These algorithms are now capable of estimating the three-dimensional coordinates of key points, further enhancing their applicability.

Advances in computer vision technology have refined the capability for precise data collection of micro-behaviors. However, methods for computationally analyzing this data to extract meaningful information remain underdeveloped. To quantify and interpret the information embedded in behavioral data, it is essential to address the following questions: Which behavioral metrics are of interest in micro-behavioral research? How can these metrics be represented and calculated theoretically and methodologically? How can the coupling relationships between these metrics and spatial (as well as temporal) dimensions be expressed? This paper aims to develop a workflow for micro-behavior computation and address these questions.

3 Micro-Behavior Computation Based on Machine Vision

3.1 Framework

Building on object detection and tracking, integrating human pose estimation algorithms to obtain human key point coordinates is a critical foundation for improving the resolution of behavior data and enabling micro-behavior analysis and computation. For instance, related studies have utilized surveillance cameras in conjunction with the OpenPose 2D human pose estimation algorithm to analyze the spatiotemporal behavior distribution of individuals in buildings. In these studies, the primary information collected remains the location of individuals in indoor spaces, with behavior differentiated based on the categories of Points of Interests (POIs) adjacent to them [13]. However, current research of this kind has not fully exploited the key point data gathered through human pose estimation methods. Specifically, it has not delved into the micro-level dimensions of behavior.

This paper proposes a micro-behavior computation framework based on machine vision (monocular cameras), which consists of two parts. The first part is behavioral data collection and preprocessing. By employing 3D Human Pose Estimation algorithms on the foundation of object detection and tracking, the spatial position information of human key points can be obtained, significantly enhancing the granularity of large-scale behavioral data collection. The second part is behavioral metrics computation. Using the three-dimensional coordinates of key points and reconstructed human models, meaningful micro-behavior metrics are analyzed and computed, yielding deeper insights into micro-behavioral information.

The behavioral data collection and preprocessing component implements the technical pipeline from monocular camera image input to the output of three-dimensional human key point coordinates by integrating several computer vision algorithms. Specifically, YOLOX is employed for object detection, while ByteTrack is used for object tracking, enabling the identification of pedestrians in video sequences and ensuring consistency in pedestrian detection across consecutive video frames [14, 15]. Building on this, HRNet is utilized to extract the 2D positions of human key points [16]. Subsequently, GAST-Net is applied to regress and generate the 3D positions of human key points [17].

The behavioral metrics computation component reconstructs the 3D coordinates of key points into spatial points on the Rhino 8 Grasshopper platform. These points are sequentially connected to form a visualized micro-level spatiotemporal behavior model of the human body. Based on the research objectives, specific behavioral metrics are identified. Using the Grasshopper platform and its built-in code editor, data is processed to achieve the quantitative representation of these metrics. The data is then visualized on the reconstructed model. To illustrate the implementation of behavioral metrics computation, gait analysis is presented as an example. Additionally, a Peri-personal Space (PPS) experimental study is used to validate the feasibility of the machine vision-based micro-behavior computation method.

3.2 Micro-Behavior Data Collection and Calibration

This study represents human posture using 17 key points (Fig. 6.1). To visualize the results and facilitate the study of the interaction between human spatiotemporal behavior and spatial environments—thereby guiding spatial design—the Grasshopper platform is employed for reconstructing human spatiotemporal postures. In Grasshopper, a comma-separated value (CSV) file containing the key point coordinates extracted from each video frame is imported. Through data cleaning and processing, the 3D coordinates of the 17 key points are organized. These coordinates are then assembled into points, which are connected sequentially, as shown in Fig. 6.1, to form the skeletal model of human posture. Similar to early experiments using continuous exposure photography to study behavior, this method produces a continuous 3D model of human postures. To make the postures visible within dense models, frame-skipping can be applied for display purposes. Additionally, the models of consecutive frames can be spaced apart for clearer observation.

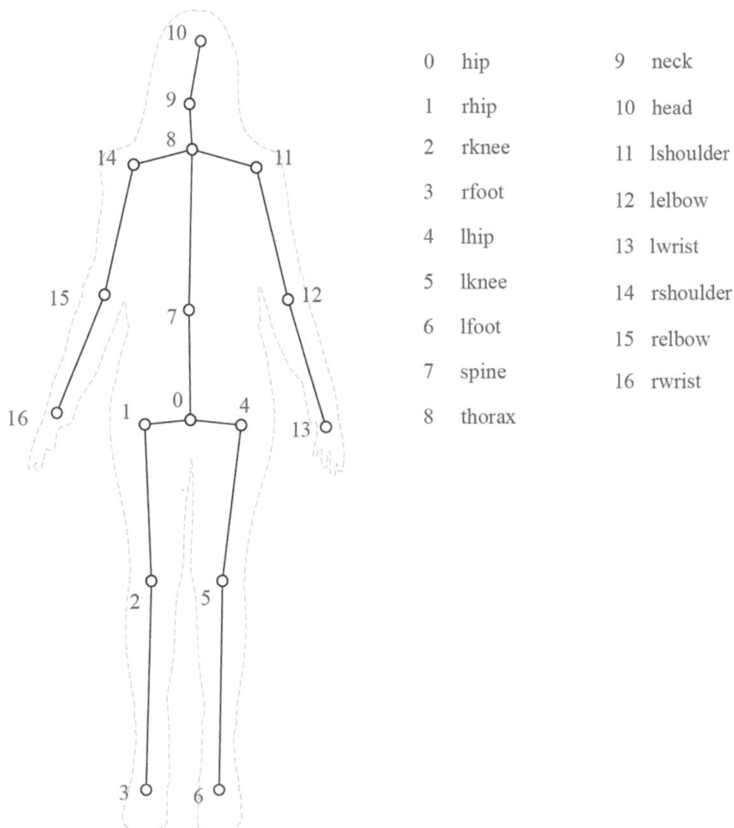

Fig. 6.1 Joint numbering and connection scheme of the skeletal model used in this study

Preliminary experiments revealed that the camera's shooting configuration affects the accuracy of human key point data collection. While monocular camera-based video estimation methods offer significant flexibility and low requirements for camera positioning and configuration, the accuracy of results is still influenced by the spatial relationship between the camera and the experimental target due to limitations in the algorithms used during data collection and preprocessing. This study evaluates the optimal shooting configurations for improving reconstruction accuracy by comparing the results obtained at different distances and camera angles during video recording. The test subject's key point 8 is positioned at the center of the camera frame, with the camera tilt angle α set to 15°, 30°, 45°, 60°, and 75° relative to the horizontal plane. The straight-line distance D between the camera and the subject is set to 3 m, 5 m, 7 m, 9 m, and 11 m (Fig. 6.2). During the experiments, configurations where the camera height H exceeded 5 m were excluded, resulting in 12 experimental conditions. The relevant parameters are listed in Table 6.1. In three configurations (45° at 3 m, 60° at 3 m, and 75° at 3 m), human posture could not be identified, leaving nine conditions with successful reconstructions. The actual lengths of the subject's leg joints (joint 4–5 and joint 5–6) were measured and compared with the reconstruction results. The findings indicate that the condition with a 15° tilt at a 5-m distance yielded the smallest root-mean-square error (RMSE). For all tests with tilt angles less than or equal to 30° and distances less than or equal to 9 m, the RMSE for joint 4–5 reconstruction was less than 13 cm, and for joint 5–6, less than 9 cm. The reconstruction errors of joint lengths can be further reduced in cases where the subject's height is known by scaling the overall height of the reconstructed model.

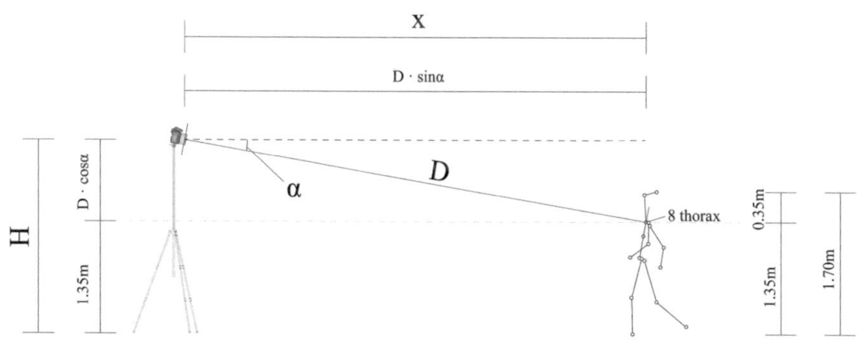

Fig. 6.2 Schematic diagram of the accuracy measurement experiment

Table 6.1 Parameters of the accuracy measurement experiment

[x(m), H(m)]	D = 3m	D = 5 m	D = 7 m	D = 9 m	D = 11 m
$\alpha = 15°$	[2.90, 2.13]	[4.83, 2.64]	[6.76, 3.16]	[8.69, 3.68]	[10.63, 4.20]
$\alpha = 30°$	[2.60, 2.85]	[4.33, 3.85]	[6.86, 4.85]	[/, 5.85]	[/, 7.85]
$\alpha = 45°$	[2.12, 3.47]	[3.54, 4.89]	[/, 6.30]	[/, 7.71]	[/, 8.13]
$\alpha = 60°$	[1.50, 3.95]	[/, 5.68]	[/, 7.41]	[/, 9.14]	[/, 10.88]

(*continued*)

Table 6.1 (*continued*)

[x(m), H(m)]	D = 3m	D = 5 m	D = 7 m	D = 9 m	D = 11 m
$\alpha = 75°$	[0.18, 4.25]	[/, 6.18]	[/, 8.11]	[/, 10.04]	[/, 11.98]

3.3 Micro-Behavior Metric Calculation and Visualization

In the software interface, the reconstructed skeletal model can display the temporal and spatial sequences of micro-behaviors. Through further behavioral calculations based on spatiotemporal coordinate points, meaningful metrics for behavior analysis can be obtained. Taking gait analysis as an example, relevant gait metrics include stride length, stride width, stride velocity, double support time, and stride velocity variability (Fig. 6.3). Strides are segmented based on the landing of the ankle joints 3 and 6. Behavioral calculations are performed using the landing point coordinates to derive the numerical values of the aforementioned gait metrics. To visualize these metrics, the gait data is mapped to different colors, which are then applied to each stride's skeletal model. This approach enables the visual presentation of gait metrics.

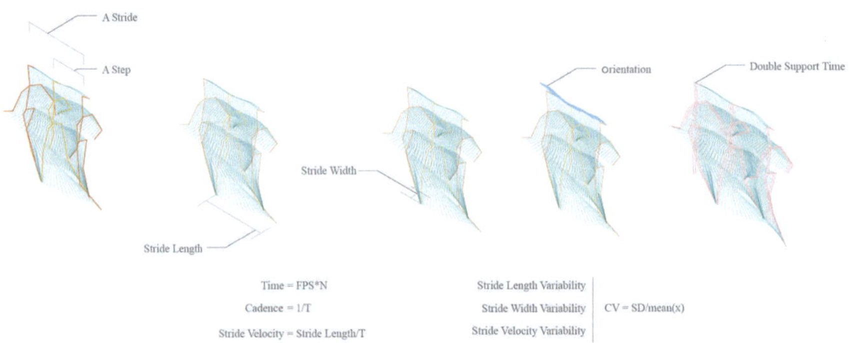

Fig. 6.3 Calculation of gait-related metrics

To verify the effectiveness of the monocular video-based micro-behavior calculation framework, this study compares gait metrics obtained through motion capture (MoCap) with those derived from the monocular calculation method (Fig. 6.4). In the experiment, reflective markers were custom-placed at the same positions as the 17 keypoints predefined in the human pose estimation algorithm. Two participants walked separately and together along a cross-shaped path within the motion capture area. The motion capture system recorded 17 joint coordinate data at 90 frames per second (fps). Simultaneously, a mobile phone was used to record video, and the monocular behavior calculation method was applied during post-processing to obtain 17 joint coordinate data at 30 fps.

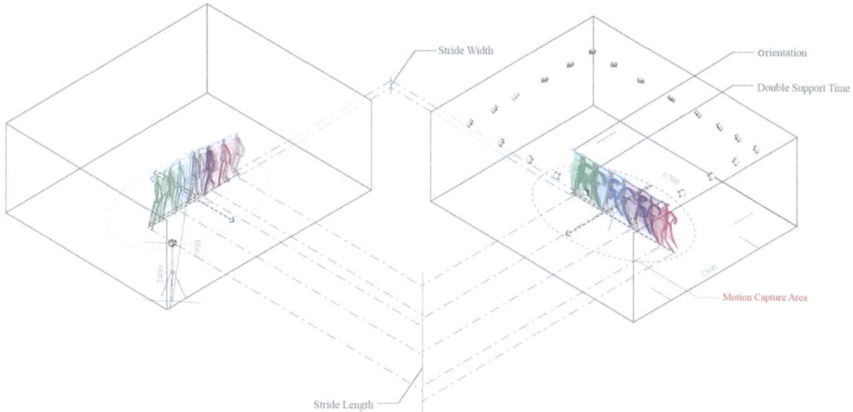

Fig. 6.4 Illustration of the motion capture experiment

The results indicate that both the motion capture and video analysis methods yield consistent outcomes in stride segmentation. After applying projection transformations, the trajectory shapes are largely similar. Based on this, two versions of gait metric data describing the same walking sequence (a selected series of four consecutive steps) were obtained. For these four steps, the RMSE of stride length was approximately 19 cm. The video analysis method effectively identifies realistic footfall points within the behavioral trajectory, enabling micro-behavior studies coupled with spatial factors, such as proxemics research. Additionally, the video analysis method can synchronize time frames of simultaneous behaviors and incorporate information on micro-behavior orientation and dwell time, providing a foundation for spatiotemporal micro-behavior studies, such as crowd close-contact research.

4 Micro-Behavior Computation Method Validation

4.1 Social Behavior and Proximal Personal Space Experiment

To validate the effectiveness of the proposed behavior computation method, a preliminary application was conducted in a social behavior study. This experiment explored the relationship between human Peri-Personal Space (PPS) and environmental factors. Obstacles and obstructing individuals were placed in the elevator lobby on the third floor of a teaching building, observing how participants exiting the elevator navigated around the obstacles. The experiment employed two macro-level indicators and one micro-level indicator to represent the PPS radius, aiming to explore patterns of avoidance distance when participants encountered obstacles (Fig. 6.5). The Turning Point in Trajectory, a macro-level indicator, was identified by comparing participants' center-of-mass trajectories with the average trajectory of pedestrians in the absence of obstacles. Another macro-level indicator, the Closest Distance, measured the shortest horizontal distance from the obstacle to the participant's center-of-mass trajectory. In contrast, the micro-level indicator, Turning Point in Gait, was derived from the geometry of footfall points, incorporating detailed micro-level information such as the sequential differences in vector magnitude between steps. Figure 6.6 illustrates how Turning Point in Gait can represent PPS. By analyzing the gait geometry, footfall points with larger vector magnitude differences between steps and shorter distances between the left and right feet were identified as turning points. These turning points, connected to the obstacle or obstructing individual, provided a precise measure of the PPS radius. The distance between the turning point and the obstacle was used to quantify the participant's judgment of personal space in the given environment.

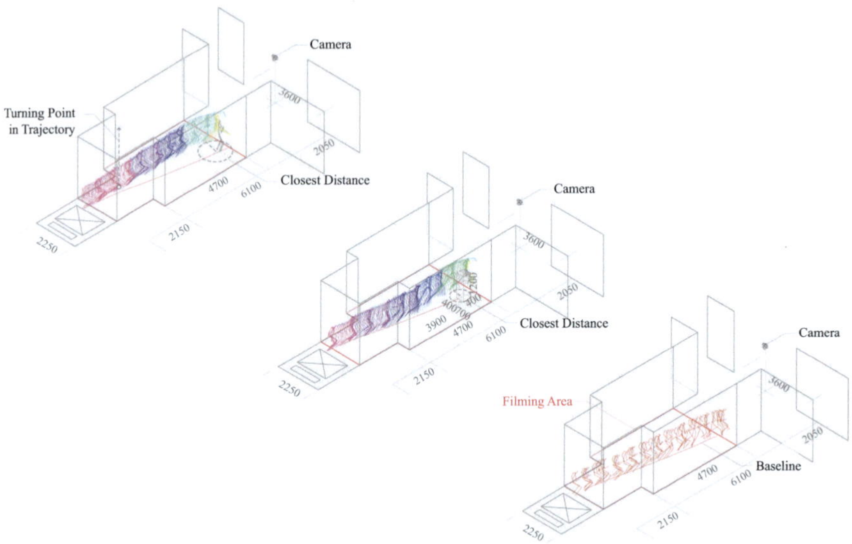

Fig. 6.5 Schematic diagram of the PPS experiment

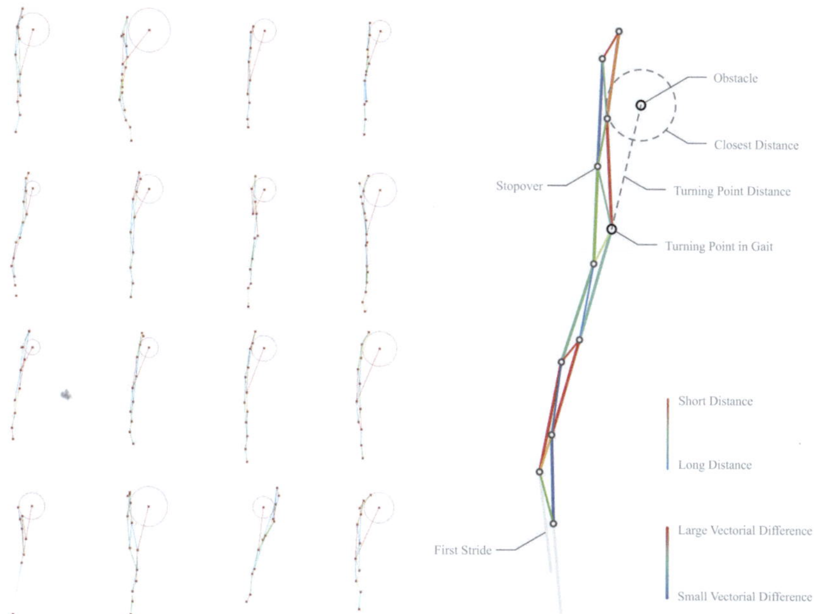

Fig. 6.6 Stride Geometry: Visualization of vector differences and turning point positions

4.2 Peri-Personal Space (PPS) Analysis Based on Micro Gait Data

The results of representing PPS radius using Turning Point in Trajectory, Closest Distance, and Turning Point in Gait exhibit slight differences. Determining Turning Points in Trajectory requires comparing against the average trajectory from a control group without obstacles. The subject's trajectory is fitted using an eighth-degree polynomial regression, and the result is subtracted from the control group's average trajectory polynomial. The point where the second derivative of the difference function reaches its maximum is identified as the Turning Point in Trajectory on the original trajectory. In some experiments, participants exiting the elevator deviated significantly from the initial direction of the control group's trajectory. In such cases, the calculated trajectory turning point often falls near the elevator entrance, which is not accurate for representing PPS. The Closest Distance, defined as the closest horizontal distance between the body's center of mass and the obstacle, does not rely on a baseline comparison. However, since individuals vary in their response to obstacles, using the nearest distance to represent PPS is influenced by individual differences. PPS signifies the distance at which a person becomes significantly aware of the presence of an obstacle. Among the three methods, the Turning Point in Gait, a micro-level metric, can exclude certain confounding factors and objectively and accurately represent the PPS radius. Based on experimental video assessments, the results obtained through micro-gait analysis align more closely with real-world conditions.

5 Discussion

This study, through methodological construction and preliminary experiments, has initially demonstrated the advantages of micro-behavior computation in analyzing complex spatial behavior patterns. However, the current technical methods still require further optimization and refinement to improve the accuracy of the analyses. For example, in the analysis of gait data, stride segmentation relies on setting a height threshold to determine the landing state of joints 3 and 6, which are then used to divide strides. In this study, the simulated annealing algorithm within the Galapagos solver on the Grasshopper platform was employed to optimize the height threshold value, aiming to maximize the number of segmented strides. While this approach is effective in most cases, it still encounters errors in a few specific instances. Consequently, further optimization of the stride segmentation criteria is needed.

Moreover, the proposed technical methods in this study need to address practical considerations of operational feasibility and convenience during experimental applications. For instance, the behavior collection and preprocessing method employs a top-down architecture, which involves detecting targets first and then estimating 3D poses within each detection box. Following this step, to determine relative positional relationships between individuals, it is necessary to apply inverse perspective transformation to the video images to obtain orthogonal spatial projections. Currently, this issue is addressed by identifying four points in the image coordinate system of the video and their corresponding points in the real-world coordinate system, generating a perspective transformation matrix, and applying it to all frames. However, due to variations in video recording setups, this process is prone to significant human error during practical operation. An alternative approach involves marking a predefined rectangular area with known side lengths in the experimental field before recording. After fixing the camera position, participants are asked to walk along the edges of this rectangle to establish a baseline reference for their motion trajectories. During video processing, the initially estimated pose coordinates (without inverse perspective transformation) are adjusted using a mapping matrix derived from the endpoints of the rectangular area. This matrix is then applied to all targets. While this method effectively reduces human error and improves data accuracy, it also introduces additional complexity to the experimental procedures.

Future research can further focus on the accuracy assessment and validation of behavior computation methods in multi-person scenarios. In such contexts, behavior analysis will extend to more complex metrics, such as modes of interaction between individuals. This advancement will provide a reliable framework for describing and characterizing group behaviors in real public spaces.

6 Conclusion

The application of machine vision-based behavior analysis methods is becoming increasingly mature in the field of architecture. This study argues that such technology not only facilitates large-scale automated behavior analysis but also significantly enhances the precision and depth of spatial behavior studies. Building on this foundation, the study proposes a micro-behavior computation method based on machine vision, capable of evaluating various micro-behavioral metrics, such as effective social distance and micro-level close-contact situations. Furthermore, it enables the visualization and reconstruction of micro-behavioral computation results within spatial models. By establishing a cognitive understanding of the correspondence between micro-behaviors and spatial geometry, this method provides a foundational tool for architects to make informed design decisions.

References

1. Song S, Xiao M, Lu X (2024) Body, behavior and space: revisiting the concept of small scale in public spaces. Urban Plan Int 39(1):60–66
2. Yang CX, Zhan M, Geng HZ, Liu MX, Lyv CZ (2022) Research trends of outdoor public space and key problems of waterfront based on behavior simulation. J Hum Settl West China 37(5):89–97
3. Xuan XD, Yan MH, Zou J, Zheng YH, Zhang ZX (2024) Exploring the influences of the tourist path space quality on visitor behaviors under a physical environment intervention: a case study based on Sanhe Ancient Town in Hefei City. South Arch 5:14–28
4. Yan B, Fu ZQ, Xu JX (2023) Public space and behavior interaction of commercial complex. Urban Arch Space 30(3):105–108
5. Li XY, Shan JY, Chong ZG (2020) A study on strategies for improving the transfer experience of underground integrated transportation hub from the perspective of stereoscopic city: the case of Beijing Sub-center station. New Arch 06:22–26
6. Fouad AA, El Antably A (2023) Agent-based modeling and simulation of pandemic propagation in a school environment. Int J Archit Comput 21(1):120–135
7. Lin YM, Huang WX (2021) Research on passengers' spatial-temporal behaviors in the airport terminal based on wi-fi positioning technology: a case study of Shanghai Hongqiao international airport. New Arch 02:24–30
8. Gu ZC, Lu W, Osaragi T (2022) Study on spatio-temporal behavior in underground pedestrian system based on choice level theory. Chin J Undergr Space Eng 18(S01):1–13
9. Jiang DH, Xia L (2014) Research and application of video stream statistical technology. Mod Arch Electr 5(1):1–4
10. Wu SJ, Hu YK (2022) Research on spatial hyper-links in commercial complexes based on deep learning: a case study of Tai Koo Li Sanlitun and Beijing APM. South Arch 1:61–68
11. Zhang ZX, Long Y, Wang MY, She YS, Zhang A, Jacobs S (2019) Application of wearable cameras in studying individual behaviors in built environments. Landsc Arch Front 7(2):22–38
12. Zhang N, Liu L, Dou ZY, Liu XY, Yang XZ, Miao DD, Guo Y, Gu SL, Li YG, Qian H, Wei JJ (2023) Close contact behaviors of university and school students in 10 indoor environments. J Hazard Mater 458:132069
13. Yang LJ, Huang WX, Wang YC (2020) Application of image recognition technology on architecture behavior research. In: Proceedings of 2020 national conference on architecture's digital technologies in education research, pp 145–151
14. Ge Z, Liu ST, Wang F, Li ZM, Sun J (2021) Yolox: exceeding yolo series in 2021. arXiv: 2107.08430
15. Zhang YF, Sun PZ, Wang XG (2022) Bytetrack: multi-object tracking by associating every detection box. In: European conference on computer vision, pp 1–21
16. Sun K, Xiao B, Wang JD (2019) Deep high-resolution representation learning for human pose estimation. In: Proceedings of the IEEE/CVF conference on computer vision and pattern recognition, pp 5693–5703

17. Liu JF, Rojas J, Li YH, Liang ZJ, Guan YS, Xi N, Zhu HF (2021) A graph attention spatio-temporal convolutional network for 3D human pose estimation in video. In: 2021 IEEE international conference on robotics and automation (ICRA 2021), pp 3374–3380

Growth and Iteration: Exploring Tectonic Design with Natural Tree Branch Materials from the Perspective of Tree Morphology

Zhixin Wen, Congcong Ren(✉), and Xin Yan

Beijing University of Civil Engineering and Architecture, Beijing, China
wenzhixin0509@163.com, {rencongcong,yanxin}@bucea.edu.cn

Abstract. The growth process of trees embodies efficient and ingenious structural characteristics, where high-strength fibers grow along the axial direction of branches, making it a biologically smart material that is difficult to simulate. For a long time, due to various factors, a large amount of unconventional lumber, such as natural tree branch materials, has not been utilized in the construction of buildings, resulting in the wastage of its unique organic structural morphology and mechanical properties. Based on the perspective of tree morphology, this study analyzes and learns from the growth patterns, structural forms, and environmental adaptations of trees. It proposes an iterative algorithm for simulating growth structures based on the Grasshopper platform, and elaborated on the issues of data transformation and control. In terms of subjective control methods, a manual adjustment method and an automatic generation method were innovatively proposed and tested to varying degrees. Regarding digital construction, the study focused on simplifying joints and innovating mortise and tenon structures, conducting preliminary construction experiments. This research initially explores a method for the design and construction of wooden structures using natural tree branch materials from the perspective of tree morphology. It is dedicated to finding new wooden structural forms that fully utilize the organic structural characteristics of trees.

Keywords: Natural wood · Tree morphology · Computational design · Design-to-fabrication · Traditional structure

1 Introduction

1.1 Background

In recent years, wood, as a traditional building material, has regained attention and undergone a revival globally. However, the mainstream wooden architecture still primarily uses standardized straight components as the core structural elements. Looking back at history, instances of unconventional wood usage, though existing, are relatively few. For instance, in 17th-century Europe, the shipbuilding industry had a history of utilizing curved wood [1] (Fig. 7.1), while in China, some non-standard wood was employed in the

J. Fu et al. (eds.), *AI for Architecture*, Computational Design and Robotic Fabrication,
https://doi.org/10.1007/978-981-95-0974-4_7

construction of southern pile-dwelling houses and northern open-timber roof structures [2]. Nevertheless, due to the historical predominance of orthogonal or beam-column systems in architecture, and the tendency for architectural volumes and structural demands to favor straight components, this limited the widespread application of unconventional wood.

Fig. 7.1 Curved wood used in European shipbuilding in the 17th century

Since modern times, influenced by the development of concrete and steel structure buildings, wooden architecture technology has also made progress, such as the development of high-quality cross-laminated timber (CLT). However, the application of unconventional wood, such as natural forked materials, is still insufficient.

1.2 Related Research

In recent years, there have been some studies on the construction utilization of unconventional wood. These studies mainly focus on the combination of structural optimization and form-finding in wooden structures with robotic digital construction. However, they lack deep exploration into the tectonic logic and show a tendency towards homogeneity. In this regard, these researches tend to first adopt general methods to design an original structure and then use specific methods and material libraries for matching. For instance, a geometric matching method, which was based on a Vierendeel-style arching truss, was implemented [3]. Additionally, a matching approach that originated from ParaGen was put to use [4]. Other research teams either adopted the Hungarian algorithm-based matching method [5] or implemented a matching technique anchored in graphic statics [6].

The essence of such practices is to make unconventional wood and topological form-finding methods adapt to each other, with the tectonic logic based on preexisting design methodologies. However, the unique characteristics of unconventional wood originate from its growth process, including its structural properties and organic forms. To better utilize this material, rather than just reducing construction waste, it is necessary to explore a tectonic logic and method that align with its growth process and leverages its structural and morphological features.

2 Analysis of Tree Growth in Nature

The growth direction and morphology of trees are the result of multiple factors, including both environmental and internal factors.

2.1 Environmental Factor: Light

Research has shown that trees adapt to their environment primarily through secondary growth processes, which determine their survivability. Among the environmental factors in nature, the photoperiod is the most stable, playing a dominant role in controlling the start, stop, and speed of tree growth processes [7]. On the side with sufficient light, tree branches tend to grow towards the light source to maximize photosynthetic efficiency. This phototropic growth results in branches forming larger angles in the direction of light, making the tree canopy more open. The size difference in the gaps formed during tree growth largely determines the success of branch growth. A larger gap provides a greater advantage, increasing the likelihood that the branch will eventually reach the canopy layer and contribute to a lush, vibrant growth [8].

2.2 Environmental Factor: Climate

In the context of global warming, temperature and climate are important factors that affect tree growth, and continuous climate change will trigger a geographical redistribution of growth responses to climatic factors [9]. Among these factors, rain and snow can cause mechanical damage to tree branches, affecting their growth direction and morphology. Therefore, in areas with more rain and snow, trees often develop more resilient and flexible branch structures to adapt to this stress. Meanwhile, wind plays a crucial role in the changes during tree regeneration and succession stages. Under strong winds, trees can reposition themselves, reconfigure their canopies, or shed leaves and branches to reduce resistance [10]. In areas with stronger winds, to reduce wind resistance and enhance stability, trees often grow lower and have smaller branch angles to minimize the impact of the wind.

2.3 Internal Factors of the Plant

From a microscopic perspective, axial growth indeed occurs along the fiber direction within the tree, and this growth pattern plays a crucial role in the tree's resistance to various natural forces, thereby maintaining its stability. The fiber direction of tree branches is primarily longitudinal, arranged parallel to the main axis of the branch. Additionally, the tensile and compressive strength of wood along the fiber direction (i.e., axial direction) is much greater than its strength under transverse (radial and tangential) forces [11]. This arrangement allows them to better distribute and resist pressure when subjected to gravity or external impacts, making the tree more stable when withstanding natural forces such as wind and snow pressure.

In summary, the various factors influencing the transformation of a sapling into its final form through the growth process have long contributed to the environmental adaptability of trees. Through this adaptability, trees can overcome survival difficulties under different conditions and form stable macro and micro structures. This project embarks on a tectonic exploration from the perspective of tree morphology. The aim of this study is to imitate and simplify the tree growth process into quantifiable geometric parameters by extracting geometric features. Utilizing these geometric parameters, we seek to simulate the tectonic logic of tree growth and control the structural form of constructions, thereby better exhibiting the inherent characteristics of unconventional lumber.

3 Iterative Algorithm for Simulating Growth Structures

Through the analysis of tree morphology, two simplified geometric features have been identified as essential for representing tree growth patterns. Firstly, the direction of extension must be axial. Secondly, the angle of growth varies to adapt to the environment.

The axial direction, as a critical component of tree growth, refers to the characteristic of trees to extend along the branch angle after each bifurcation. In the absence of human intervention, such as pruning or suppression, most tree branches maintain a relatively stable extension angle before undergoing the next bifurcation, without sudden angle changes. This characteristic plays a decisive role in the unique mechanical properties of the tree's fiber structure. Preserving the axial growth logic is particularly important when utilizing unconventional lumber to construct structures that can fully leverage its performance. This viewpoint has been widely recognized in previous related studies [4, 5].

The angle of branch development, which determines the direction of each bifurcation, influences the growth trend before the next bifurcation. The core of the simulated growth algorithm lies in transforming the natural variation of tree growth angles into easily controllable algorithmic parameters. Unlike previous studies that initially searched for shapes and then matched them, this study approaches the problem from the perspective of existing unconventional lumber objects and constructs based on a digital material library, utilizing growth iteration logic.

There are mainly two challenges in this approach. The first challenge is how to extract the angles from existing unconventional lumber for iteration, rather than for matching (*Data Conversion Issues in Existing Material Libraries*). The second challenge is determining the method to control the direction of each iteration (*Control Issues in Iterative Algorithms*). The following are two issues elaborated separately.

3.1 Data Conversion Issues in Existing Material Libraries

In projects involving construction with existing components, material database establishment issues in material libraries are quite common, and relevant research provides valuable references [5, 6]. Initially, discarded tree branches were collected from the horticultural departments of Beijing Zoo and Zizhuyuan Park after pruning, which served as raw materials. The excess parts of them were then trimmed at the Beijing University of Civil Engineering and Architecture's woodworking workshop. Following this, 3D scanning technology was employed to digitally model these branches. The Mesh models underwent post-processing, were centrally numbered, and a database was established (see Fig. 7.2).

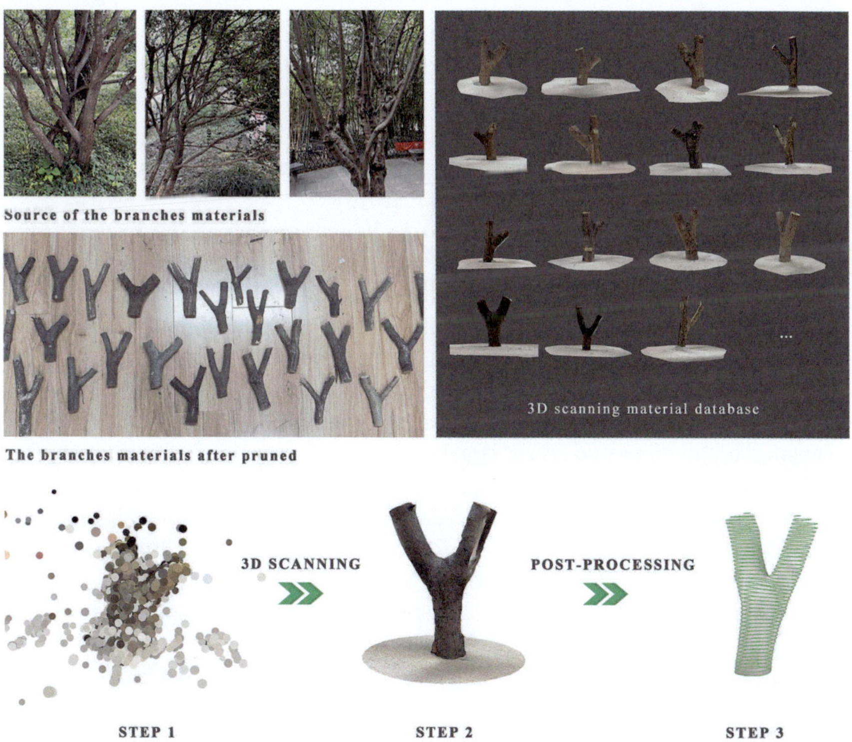

Fig. 7.2 Data conversion issues in existing material libraries (Including the whole process from material collection to 3D scanning)

To facilitate the extraction of required geometric data from the material library, Rhino's Grasshopper parametric design platform was utilized. By analyzing the center points of the cross-sections of tree branches, the complex material models were simplified into linear models. This approach was derived from an understanding of wood cross-sections. The simplified linear models contain ‘n’ length data points and ‘n + 1’ angle data points. Additional angle data was obtained by setting a special plane with its origin at the bifurcation point.

3.2 Control Issues in Iterative Algorithms

Subsequently, based on the constructive logic of growth iteration and the simplified linear model, we designed a geometric variation method that simulates growth for data conversion in the material library. This method also serves as the core of the iterative algorithm, influencing its control logic (see Fig. 7.3).

This approach consists of a three-step process:

1. Using the line extended from the previous iteration as Line 1, perform the first rotation on a specific plane by an angle of ‘a’ to obtain Line 2.
2. Rotate Line 1 on the same plane for a second time by an angle of ‘b’ to get Line 3.
3. Rotate Line 3 on a plane perpendicular to the original plane by an angle of ‘c’ to generate Line 4.

Ultimately, a new tree branch is formed by Lines 1, 2, and 4. The entire process can be fully represented by just three angle data points: a, b, and c. This also implies that with these three data points, both data conversion in the material library and control of the iterative algorithm can be achieved (see Fig. 7.3).

Using this method, a large number of tree branch models in the material library can be converted into sets of three data ports, including a switch to adjust the positive or negative connection of the branches. It should be emphasized here that adjusting the positive and negative orientation of tree branches involves remodeling the tectonic logic. On one hand, pure biomimicry is disrupted, as the iterative algorithm does not merely imitate tree growth; on the other hand, it facilitates designers to create new forms. In subsequent research, the algorithm demonstrates that regular combinations of positive and negative orientations can create structural forms with principles of spaced growth and logical order.

During the algorithm’s execution, the data ports of the tree branch models are inputted into multiple steps of the algorithm in the form of GH components. Ultimately, controlling the growth iteration algorithm can be accomplished by simply changing the number of each newly grown tree branch and adjusting its axial rotation angle. Additionally, a method is provided to select the branches that need to continue growing after each iteration, allowing the rest to stop growing and preventing uncontrollable data situations (see Fig. 7.4).

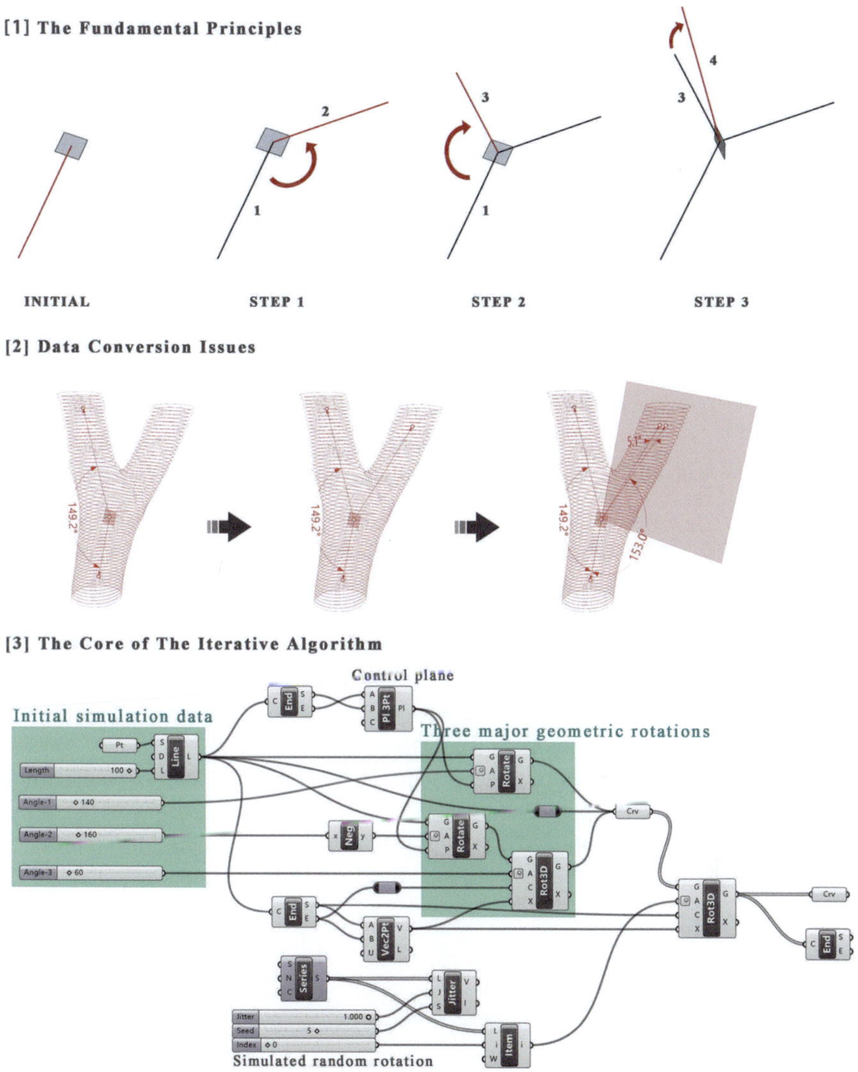

Fig. 7.3 The geometric variation method applied to both data conversion and control issues in the iterative algorithm

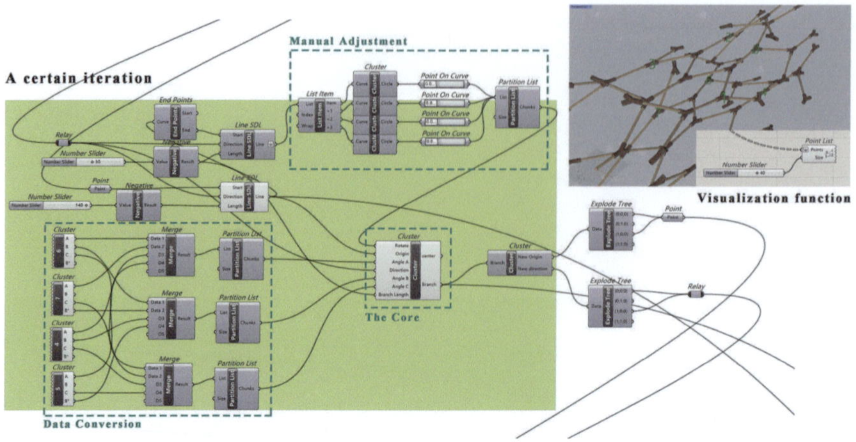

Fig. 7.4 Iterative algorithm process and visualization capabilities based on the GH platform

4 Subjective Control Methods and Construction Exploration by Designers

After completing the design of the simulated growth iteration algorithm, its tremendous potential in construction has been recognized, representing a unique design approach and tectonic logic. Initially, through initial experiments, such as adjusting axial rotation and directional changes, various structural forms have been created (see Fig. 7.5). Although these structures meet certain usage requirements and have passed structural calculations, they cannot be effectively controlled, and there are some contradictions in the details. Expecting them to be better utilized by designers or to create more fascinating structural trends necessitates special control methods to meet subjective demands.

Fig. 7.5 Structural forms of three early tests

In this regard, some preliminary exploratory research has been conducted. Two potential solutions are proposed: Manual adjustment and Automatic generation: *Manual adjustment* and *Automatic generation.* Among them, this article focuses on elaborating manual adjustment. The manual adjustment method allows designers to make detailed adjustments based on personal needs, enabling flexible control over the morphological trends of different stages and parts during the growth process, thus achieving free manipulation of the entire growth form. However, this method is more suitable for smaller-scale branch structures. In contrast, the automatic generation method needs to be combined with other algorithms to control morphology by finding optimal data that meets specific requirements, making it more advantageous when dealing with larger-scale branch structures, this method is still in the initial exploration stage.

Based on the simulated growth iteration algorithm, adjustments have been made to enable precise positioning of each branch through labeled serial numbers, allowing individual rotation and directional changes. Designers can gradually control the shape of the structure according to the iteration sequence. This regulation method, known as the manual adjustment method, is suitable for integration with AR-assisted construction technology. This approach enables designers to experience the construction process in a tactile manner and closely observe the geometric transformation of each branch.

The study has conducted multiple tests on this method and successfully designed a structural unit. This unit can be presented as an independent installation art or integrated into a larger-scale architectural structure as a crucial component of a column. To verify its structural stability and practicality, structural tests were conducted, and a 1:50 scale 3D-printed model was further produced to visually demonstrate its final presentation (see Fig. 7.6).

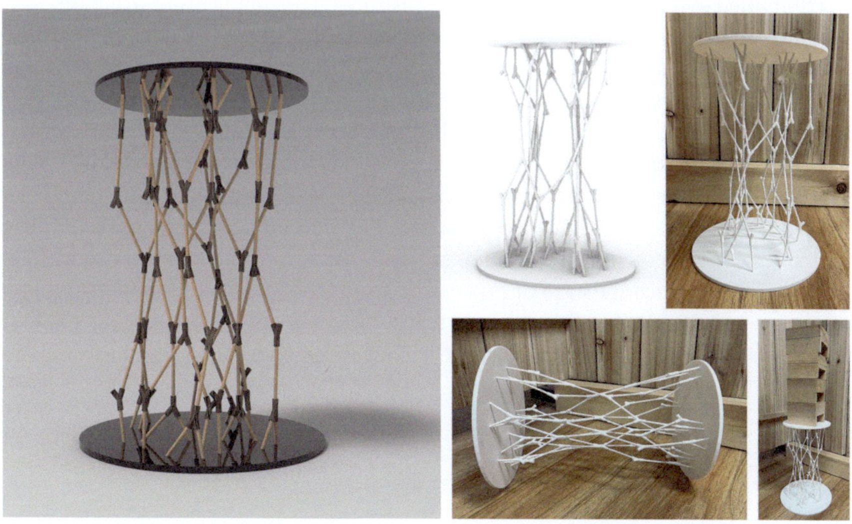

Fig. 7.6 Manual adjustment case and a 1:50 scale 3D-printed model

The other regulation method is the automatic generation approach, with the aim of enabling the growth iteration algorithm to extend along a specific surface to achieve a particular form. Currently, the simulated annealing algorithm has been adopted for initial testing. This algorithm can locate the closest point to the surface during each iteration, assisting numerous branches to precisely pinpoint the most suitable position on the surface. Nevertheless, certain contradictions and issues arise during the actual operation process, necessitating continuous optimization and enhancement of this advanced algorithm in future research.

5 Construction Attempts of Mortise and Tenon Joints in Timber Structures

In the design process of timber structures using natural tree branches, the exploration of joint design and construction methods is also crucial. In the previous section, the paper has elaborated on the value of unconventional lumber as a construction material from the perspective of tree morphology, as well as the construction logic that can maximize these values and advantages. In fact, besides a generative algorithm that can assist design, matching joints and construction methods are also an indispensable part of realizing its value.

In past related studies, most research teams have focused on utilizing digital manufacturing techniques to process unconventional lumber, such as milling operations performed by robotic arms, this is primarily due to the complex shapes and difficult positioning of unconventional lumber. However, the high capital costs and technical thresholds make it difficult for ordinary designers to access and apply. Our team has been committed to modernizing traditional construction techniques and structural forms, integrating them as a lubricant with digital fabrication to address transitional issues in current construction activities [12, 13].

This study focuses on enhancing the performance of unconventional lumber by optimizing wooden joints while simplifying the complexity of digital fabrication. In traditional Chinese mortise and tenon structures, straight tenon, dovetail tenon, and lap joint tenon each have their advantages and disadvantages. After conducting numerous tests in our workshop specifically for the joint areas of tree branches, we have ultimately designed a new type of mortise and tenon joint (see Fig. 7.7). This new mortise and tenon joint incorporates the ease of processing from the lap joint tenon while retaining the elegance and simplicity of the straight tenon. Not only does it enhance the quality of the finished structure, but it also reduces the technical demands during digital fabrication.

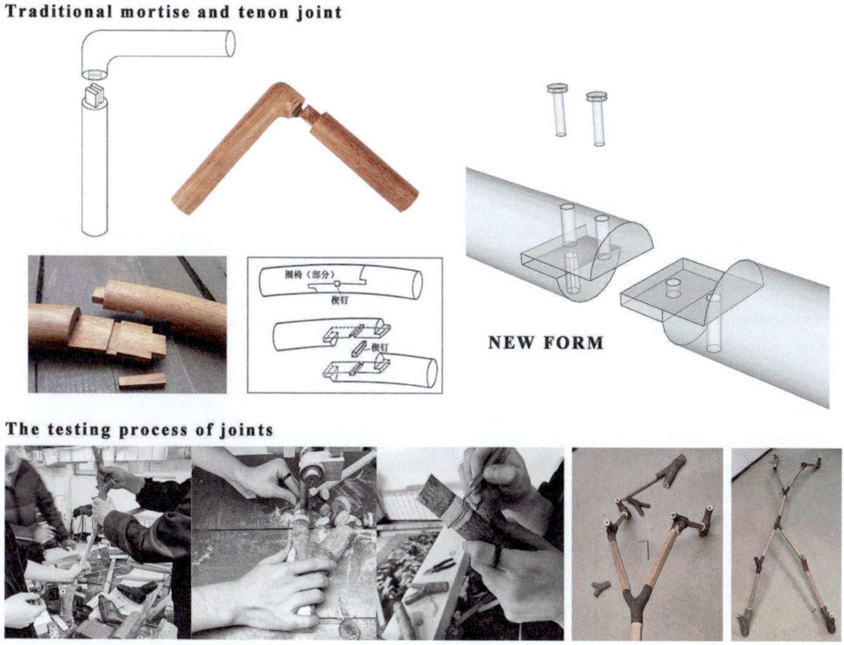

Fig. 7.7 The design and manufacturing experiment process of translating traditional mortise and tenon structures

6 Conclusion

From the perspective of tree morphology, this study conducted an exploration of the impact of environmental factors on tree morphology, as well as the growth logic and structural characteristics of tree forms, through the analysis of natural tree growth. Based on this understanding, an iterative algorithm was proposed to simulate growth structures, and elaboration was provided on the data conversion of existing material libraries and the control issues of the iterative algorithm. In terms of subjective control methods, a manual adjustment method and an automatic generation method were innovatively proposed and tested to varying degrees. Regarding digital construction, the study focused on simplifying joints and innovating mortise and tenon structures, conducting preliminary construction experiments.

Through the aforementioned research work, the powerful potential of unconventional lumber in future construction has been further recognized. This study not only provides a new perspective and approach for the use of unconventional lumber in construction but also offers designers a flexible and creative simulation method during the creative process. However, the current research still has some limitations. For instance, the accuracy of the advanced algorithm for automatic generation needs further improvement. Simultaneously, further advancement is needed in the translation of traditional joints and the construction practices of structures. These efforts will continue in future research. Ultimately, it is hoped that through these studies, more comprehensive theoretical support and technical means can be provided for the design and construction of wooden structures using natural tree branch materials.

References

1. Vial du Clairbois HS (1783) Encyclopédie méthodique marine. Panckouke, Paris, p 1783
2. Yang C (2016) Analysis on the characteristics of ancient chinese architecture. Cultural Relics Publishing House, Beijing
3. Mollica Z, Self M (2016) Tree fork truss. Adv Arch Geom 2016:138–153
4. Von Buelow P, Oliyan Torghabehi O, Mankouche S, Vliet K (2018) Combining parametric form generation and design exploration to produce a wooden reticulated shell using natural tree crotches. Proc IASS Annu Symp 2018(20):1–8. International Association for Shell and Spatial Structures (IASS)
5. Amtsberg F, Huang Y, Marshall D, Gata KM, Mueller C, Amtsberg F, Mueller C (2020) Structural upcycling: matching digital and natural geometry. Adv Arch Geom
6. Chai H, Zhou X, Gao X, Yang Q, Zhou Y, Yuan PF (2024) Integrated workflow for cooperative robotic fabrication of natural tree fork structures. Autom Constr 165:105524
7. Saderi S, Rathgeber CB, Rozenberg P, Fournier M (2019) Phenology of wood formation in larch (Larix decidua Mill.) trees growing along a 1000-m elevation gradient in the French Southern Alps. Ann For Sci 76:1–17
8. Poorter L (1999) Growth responses of 15 rain-forest tree species to a light gradient: the relative importance of morphological and physiological traits. Funct Ecol 396–410
9. Babst F, Bouriaud O, Poulter B, Trouet V, Girardin MP, Frank DC (2019) Twentieth century redistribution in climatic drivers of global tree growth. Sci Adv 5(1):eaat4313
10. Gardiner B, Berry P, Moulia B (2016) Wind impacts on plant growth, mechanics and damage. Plant Sci 245:94–118
11. Hoadley RB (2000) Understanding wood: a craftsman's guide to wood technology. Taunton press
12. Wen Z, Yan X, Ren C (2024) Digital translation of traditional tectonics: a study on parametric design and construction of mortise and Tenon Timber Pavilion. South Arch 05:50–57
13. Yan X, Bao DW, Ren C, Xie YM (2023) Constructing topologically optimized spatial structure using innovative mortise-and-tenon joints. Proc IASS Annu Symp 2023(19):1–12. International Association for Shell and Spatial Structures (IASS)

Performance-Driven Structure in AIGC: An AI-Driven Workflow of Bamboo Chair Design and Fabrication

Shitong Wang, Yulu Zhang, Yige Liu(✉), and Philip F. Yuan(✉)

CAUP Tongji University, Shanghai, China
philipyuan007@tongji.edu.cn

Abstract. With the rapid development of artificial intelligence in recent years, the use of generative design to assist in design decision-making has become an important topic. Bamboo, as a low-carbon and resilient material, is widely applied in architecture and furniture manufacturing. This study explores iterative generative design based on the bending properties of bamboo, combining performance analysis software to optimize the structure and plan the manufacturing process. The entire workflow from design to manufacturing is completed. The results demonstrate that the integration of artificial intelligence with performance-based analysis enables the seamless combination of computational innovation in design and practical construction, facilitating AI's full involvement in the entire process from design to construction. This approach provides new insights into the innovation of generative design for bamboo materials.

Keywords: Bamboo chair · Artificial intelligence · Generative design · Structural performance-based design · The entire design and manufacturing process

1 Introduction

As an important forest resource, bamboo has a large biomass and is widely used in daily human life. Its short growth cycle, high tensile strength, lightweight, and green sustainability make it a crucial component of structural materials [1]. In the context of global warming, transitioning from non-renewable and highly industrialized materials to natural, sustainable, and renewable materials is a necessary condition for achieving balanced sustainable development. As a result, bamboo has become a significant choice for future sustainable materials [2].

In recent years, with the rapid development of artificial intelligence technology, data-driven generative design has played an increasingly important role in design decision-making [3]. Through generative design software, designers can quickly create a series of designs. However, during the actual production process, AI-generated designs may face issues such as structural deficiencies or disordered topological relationships, requiring designers to participate in the generative design process for selection and modification

J. Fu et al. (eds.), *AI for Architecture*, Computational Design and Robotic Fabrication,
https://doi.org/10.1007/978-981-95-0974-4_8

[4]. In this context, Yuan Feng and others proposed a new process in 2022, titled "Intention—Forming—Simulation—Iteration—Optimization—Manufacturing [5]." This process integrates generative design with performance-based analysis, enabling artificial intelligence to fully participate in the design and manufacturing processes, offering new ideas and methods for design innovation. Digital tools empower design, becoming an important source of inspiration for designers.

This study explores a new path for the design and manufacturing of bamboo chairs, based on the bending properties of bamboo, by combining generative design with performance optimization. In terms of generative design, this research relies on the FUGenerator platform. For performance analysis and optimization, the study focuses on optimizing the bending support and base structure of the bamboo chair, employing an active bending structure to achieve curved forms, reducing the need for complex connectors, and optimizing the overall shape. In terms of assembly and installation, the chair is assembled through robotic arm collaboration. In conclusion, this study, by integrating artificial intelligence with performance-based analysis, not only achieves full-process innovation from design to manufacturing but also provides new perspectives and methods for bamboo material design. The results of this research demonstrate the immense potential of artificial intelligence in the field of design, offering a solid theoretical foundation and practical experience for future design innovations.

2 Methodology

This study explores a new path for the design and manufacturing of bamboo chairs by integrating generative design, performance optimization, and human–machine collaborative production methods. The generative design aspect is based on the FUGenerator platform, which utilizes a variety of algorithms, including the Diffusion Model, CLIP, and GAN, and supports small model training and multiple generative applications. The platform's advantage lies in assisting designers with mainstream large models to train smaller models, while allowing interaction between the designer and the platform to control the generative outcomes, making the process controllable and recognizable. In terms of performance analysis and optimization, this study focuses on optimizing the bamboo bending support structure, employing an active bending structure to achieve curved forms, reducing the need for complex connectors, and optimizing the overall shape.

2.1 Generative Design

The generative design aspect of this study relies on the FUGenerator platform, which is based on various algorithmic models, including the Diffusion Model, CLIP, and GAN. The platform supports multiple applications such as small model training, text-to-image, image-to-image, image-to-model, and controlled generation. FUGenerator is designed around the workflow of the designer and supports continuous optimization of a single image. Compared to other mainstream image generation platforms, FUGenerator offers several advantages:

1. It uses large models to assist designers in training smaller models, resulting in generated outcomes that are highly recognizable.
2. Designers interact with the platform to drive the AI design iteration process, ensuring that the generated results are controllable. Throughout this process, AI and architects inspire each other, iteratively optimizing the generation goals to adapt to different task requirements.

Training Set Annotation

To ensure that the generated results align with the intended curvature, the FUGenerator small model requires directional annotations, integrating the bending form of bamboo with the shape of the bamboo chair. Specifically, multiple small training sets need to be created, which involve step-by-step, directed modifications and optimizations of the same intended image, from the bending state of bamboo to the typical shape of a bamboo chair. These annotated data sets enhance the precision and stability of the generated images.

Step-By-Step Iteration

FUGenerator adopts a cyclical strategy of "training—model inference—generation—result optimization—training" in its interactive process. After generating an image using AI, users adjust the semantics based on their judgment of the results, re-inputting the previous generation outcome into the AI for model inference, thus completing one iteration. In each step of the iteration, designers can modify the results according to their personal preferences. Additionally, the FUGenerator divides the image output process into multiple nodes, with each node being traceable. Designers can trace each node, allowing for more precise control over the final generated image [6].

2.2 Structural Performance Analysis and Optimization

Bamboo, due to its natural strength and resilience, is an ideal material for creating curved furniture forms. These curved shapes can be achieved through active bending structures. Active bending structures are defined as lightweight structural elements that use the elastic deformation of linear or planar components to achieve curved geometries [7]. In bamboo chair design, utilizing active bending structures can reduce the need for complex connectors and fastening devices, thus simplifying the manufacturing process. This structural design needs to consider the bamboo's elastic modulus and yield strength to ensure that no breakage occurs during the loading process.

As a composite material, bamboo's deformation behavior under different bending radii and support point locations must be determined through numerical simulation and experimental validation. Traditional shape-finding methods cannot directly obtain the balanced state of the shape, requiring lengthy and inefficient iterative attempts for precise shape control [8]. In this study, a form-driven approach is used for performance analysis and optimization, with the final form from the generative design acting as the input parameter. The absolute value of the bending moment is used as the output, and simulations are performed using bamboo strips with specific bending radii to obtain their stress distribution and the location of the peak bending moment. By adjusting the position and number of support points, the overall shape and structural performance of the bamboo chair can be optimized.

In the form-driven method, the initial target bending curve is discretized. At the equilibrium state, all mechanical forces are applied to the nodes, and the bending moment generated by the curvature is transformed into a pair of shear forces acting on the node. These shear forces are defined as [9]:

$$S_{i,i-1} = (EI)_i \frac{2 \sin \alpha_i}{l_{i-1} \cdot l_{i-1,i+1}}; S_{i,i+1} = (EI)_i \frac{2 \sin \alpha_i}{l_i \cdot l_{i-1,i+1}} \quad (8.1)$$

Variable Description (Fig. 8.1):

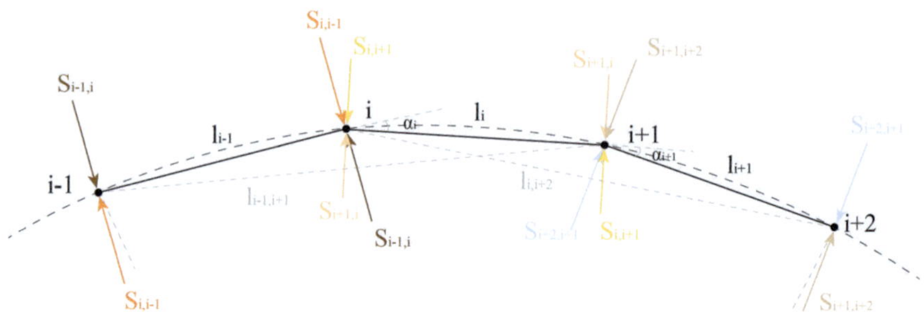

Fig. 8.1 Bending moment diagram

$S_{i,i-1}$ and $S_{i,i+1}$ represents the magnitudes of the two shear forces applied at node *i*.

α_i represents the complementary angle formed by the two straight lines adjacent to node *i*.

l_{i-1} represents the length of the line between node *i−1* and node *i*.

$l_{i-1,i+1}$ represents the length of the straight line between node *i−1* and node *i + 1*.

$(EI)_i$ represents the bending stiffness at node *i*.

After achieving equilibrium in the force diagram, the absolute values of the bending moments at each point can be calculated under the given load.

2.3 Human–Machine Collaboration

In the manufacturing process, accurately achieving the precise curvature of bamboo is a significant challenge. Traditional manual operations struggle to meet the requirements of high precision and consistency. Therefore, this study adopts a human–machine collaboration model, where designers work alongside robotic arms. This collaboration not only ensures that the curvature of the bamboo at every stage of production aligns with the expected design, but also improves manufacturing efficiency and product quality [10].

In this study, the robotic arm is employed in the key steps of bamboo bending. First, the designer creates an initial design using generative design software and utilizes performance optimization software for analysis and adjustment to determine the optimal bending path and shape. These design parameters are then input into the robotic arm's control system. The robotic arm, based on these parameters, precisely bends the bamboo, ensuring that the curvature and position of each bending point align with the design. During the bamboo bending process, human assistance is provided for tasks such as fixing and installing, while also adjusting the next step based on the initial bending results produced by the robotic arm. This approach achieves a seamless connection between design and manufacturing.

3 Case Study

3.1 Parallel Strategy of Generative Design and Performance Analysis

The parallel strategy of generative design and performance analysis is crucial for precisely controlling the design to meet both the design requirements and structural performance. This study proposes a dual-process approach where generative design and structural performance optimization work simultaneously (Fig. 8.2).

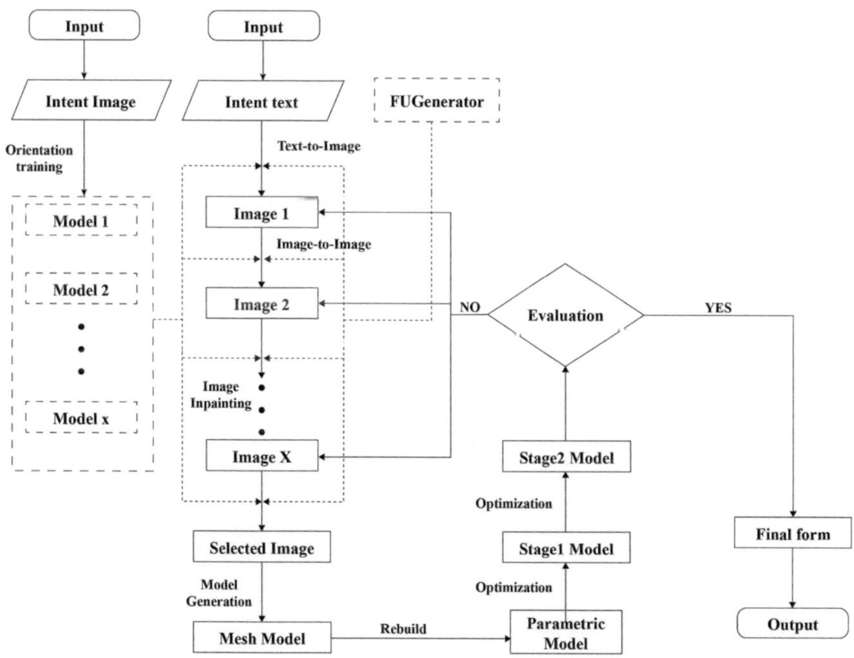

Fig. 8.2 Flowchart

Generative Design Scheme

The key to generative design lies in the directional annotation of small training sets and the step-by-step iterative design process. This study proposes a step-by-step generative design scheme, dividing the image generation into four main steps, each corresponding to a small model. This process gradually generates the final design image and mesh model. Specifically, the steps are as follows:

1. Starting from the bending form of bamboo, the "bamboo" bending shape in the wind is selected as an additional model for semantic generation. This ensures that the initial shape has a natural and harmonious aesthetic, resulting in a basic curved form that serves as the foundation for the subsequent image-to-image generation.
2. Using the basic curved form generated in Step 1, the "bamboo chair" small model is applied to generate an image of a bamboo-style chair using an image-to-image method. The focus of this step is to combine the curved shape with the structural design of the chair, ensuring that the generated image has both the natural beauty of bamboo and the functional requirements of a chair.
3. Based on the image generated in Step 2, the "material texture" small model is added. Through localized modifications, an intended image is created. This process primarily adds bamboo strip-style textures to the chair image, enriching the design with more details and realism.
4. In the final step, the intended image is refined by adding the "bamboo chair structure" small model to complete the chair's structural design details. This ensures that the structure is reasonable and feasible for actual construction, resulting in the final image.

After obtaining the final image, the FUGenerator model is used to generate a functional mesh model. Based on the mesh model parameters, further modeling of the design outcome is carried out. The flow of the generative plan is shown in Fig. 8.3.

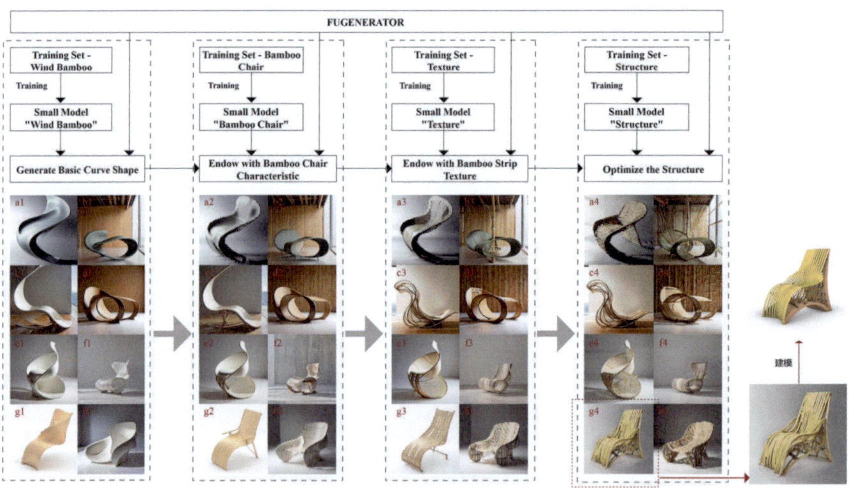

Fig. 8.3 Generative design flowchart

Performance Analysis and Optimization

This study refined the generative design scheme, selecting the g4 image as the foundation for further development. During the modeling process, we divided the design into two parts: the seat surface and the chair support structure. The seat surface is composed of thirteen curved bamboo strips, while the support structure consists mainly of round bamboo rods. Bamboo strips, being a homogeneous material, exhibit bending behavior that depends on the positioning of the support points. However, the support structure in the existing model is overly complex and not conducive to easy installation, requiring optimization.

Initially, the bending behavior of bamboo strips was studied under ideal conditions. Using Bamboo Strip 7 as an example, the simulation results, as shown in the figure, indicate that when the left endpoint of the flattened curve is set as the zero point, two peaks of the bending moment are observed at distances of 387.5 mm and 1138.7 mm from the zero point. Using the same method, we obtained the bending moment peak data for all bamboo strips (Fig. 8.4).

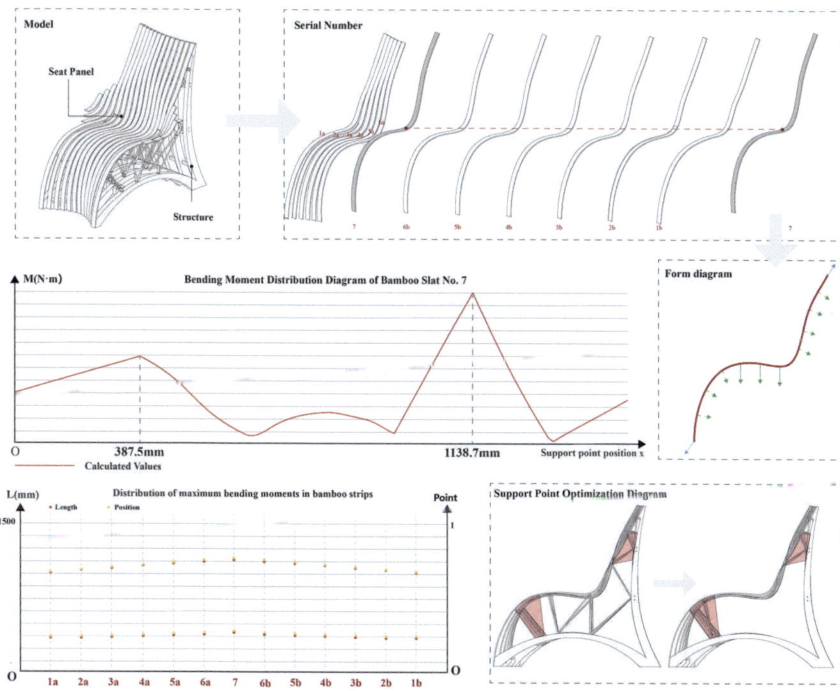

Fig. 8.4 Performance optimization process flowchart

In the optimization process, support points were concentrated at the locations of the bending moment peaks based on the simulation data. This approach not only effectively supports the bamboo strips but also simplifies the support structure, improving both installation ease and overall structural stability. The optimized cross-section results are illustrated in the figure.

Modeling, Modification, and Evaluation

Based on the performance-optimized cross-sectional results, the optimized cross-section is reintroduced into the generative design process for a new round of optimization and modification. The iterative process of modeling—modification—evaluation is repeated, ultimately resulting in the optimized support solution as shown in Fig. 8.5.

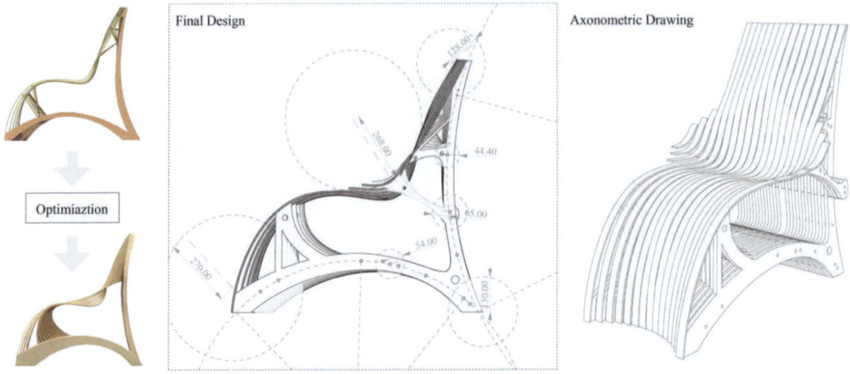

Fig. 8.5 Final design drawing

The entire process achieved the following three optimizations. First, the support structure was optimized by transforming the cylindrical rod-shaped supports into an integrated structure with the base, enhancing the overall streamlined aesthetic. Second, the bamboo strips on the seat surface were curved into arcs, employing a bending method better suited to the material's properties. Third, the base structure was optimized by reinforcing the primary stress points at the two segments of the base, ensuring the stability of the chair.

3.2 Human–Machine Collaborative Assembly Process

Assembly Sequence

This study adopts a human–machine collaborative assembly method, utilizing an ABB IRB400 robotic arm for the assembly process. The assembly sequence is shown in the Fig. 8.6. The steps are as follows:

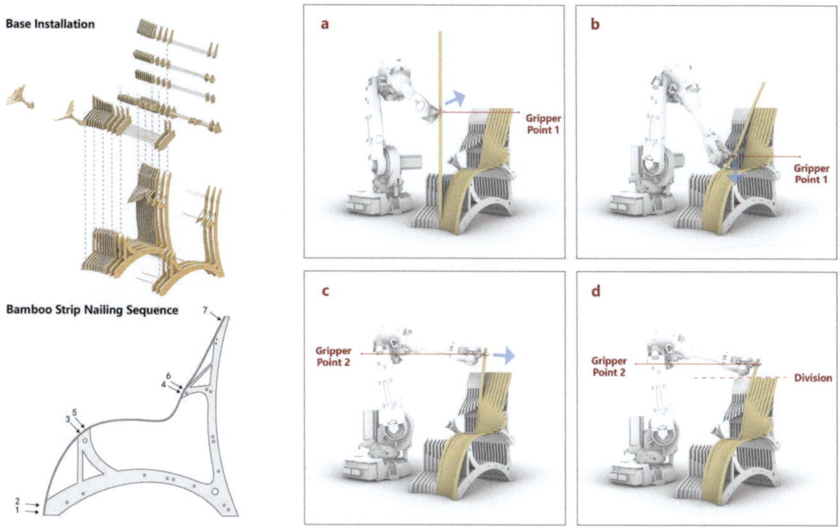

Fig. 8.6 Human–machine collaborative assembly diagram

Step 1: Begin by assembling the base structure.

Step 2: Fixing Bamboo Strips 1 and 2 at two points on the bottom of the base, with the robotic arm gripping them.

Step 3: Use the robotic arm to bend the bamboo strips downward while fixing Point 3. The robotic arm continues to press down until it reaches the design position and fixes Point 4. The robotic arm then runs the bamboo strip bending path program, adjusting the bamboo strip and fixing Point 5

Step 4: Withdraw the robotic arm and move the gripper to the end of the bamboo strip. The robotic arm brings the strip closer to the backrest until the bamboo strip is fully aligned with the base. Fix Points 6 and 7, and then cut off the excess bamboo strip to complete the installation of one bamboo strip.

The same robotic arm operation method is then used to install the remaining bamboo strips, completing the overall assembly of the bamboo chair (Fig. 8.6).

3.3 Final Result

In this study, by combining generative design with performance optimization, a complete bamboo chair production process from design to manufacturing was successfully realized. The final design and manufactured bamboo chair achieved the expected goals in both appearance and functionality. The completed bamboo chair assembly features a tightly connected base and bamboo strips, with the curved form maintained intact. The physical result matches the anticipated outcome (Fig. 8.7).

Fig. 8.7 Bamboo chair photos

4 Conclusion

This study innovatively realized the full process of bamboo chair design and manufacturing by combining generative design with performance optimization, demonstrating the immense potential of artificial intelligence in the design field. Through this research, we found that in terms of generative design, the application of generative design platforms and performance analysis tools not only significantly shortened the design cycle but also improved the accuracy and controllability of the design. Designers were able to quickly generate a series of distinguishable design solutions and optimize the generation results through interactive control. The innovation of this process lies in combining the creativity of designers with the computational power of AI, achieving efficient iteration and optimization in design.

In terms of performance optimization, numerical simulations and experimental verification ensured that the deformation behavior of bamboo at different bending radii and support positions met expectations, guaranteeing the feasibility of the design. The geometric shape and support layout of the bamboo chair base were adjusted, balancing structural strength and material usage. This not only improved overall stability but also enhanced the aesthetic appeal of the design.

During the manufacturing process, a human–machine collaboration model was adopted, where designers and robotic arms worked together. The robotic arm, through its precise control system, ensured that each bending point's curvature and position of the bamboo were consistent with the design, achieving high precision and consistency. Human–machine collaboration not only improved manufacturing efficiency but also enhanced product quality, achieving a seamless connection from design to manufacturing.

In conclusion, this study not only showcases the broad application prospects of bamboo in furniture design but also provides an effective practical case for the combination of generative design and performance optimization. Future research could further explore the optimization of more materials and structures, continuously pushing the development of design innovation, and providing new solutions for achieving more sustainable design and manufacturing.

References

1. Pothaya S, Poochai C, Tammanoon N, Chuminjak Y, Kongthong T, Lomas T, Tuantranont A (2024) Bamboo-derived hard carbon/carbon nanotube composites as anode material for long-life sodium-ion batteries with high charge/discharge capacities. Rare Metals 43(01):124–137
2. Chaowana K, Wisadsatorn S, Chaowana P (2021) Bamboo as a sustainable building material—culm characteristics and properties. Sustainability 13(13):7376. https://doi.org/10.3390/su13137376
3. Krish S (2011) A practical generative design method. Comput Aided Des 43(1):88–100. https://doi.org/10.1016/j.cad.2010.08.008
4. Saadi JI, Yang MC (2023) Generative design: Reframing the role of the designer in the early-stage design process. J Mech Des 145(4):041411. https://doi.org/10.1115/1.4056589
5. Yuan PF, Xu X, Li K (2022) Speculative architecture in the Anthropocene: The future of digital architecture. Architect J 09:12–18
6. Gu S, Wang R, Wu Y et al. (2023) Exploration of AI-inspired architectural generative design process based on the FUGenerator platform. In: Proceedings of the 2023 national conference on digital technology teaching and research in architectural departments and DADA2023 digital architecture symposium, Xiangtan, China
7. Lienhard J, Knippers J (2014) Bending-active structures. University of Stuttgart, Research Report, Institute for Structural Design and Structural Engineering, p 36
8. Boulic L, Schwartz J (2018) Design strategies of hybrid bending-active systems based on graphic statics and a constrained force density method. In: Proceedings of the IASS annual symposia. International Association for Shell and Spatial Structures (IASS)
9. Boulic L, Schwartz J (2017) Graphic statics principles for the design of actively bent elements shaped with restraining systems. In: Proceedings of the IASS annual symposia. International Association for Shell and Spatial Structures (IASS)
10. Petersen KH, Napp N, Stuart-Smith R et al (2019) A review of collective robotic construction. Sci Robot 4(28):eaau8479. https://doi.org/10.1126/scirobotics.aau8479

“AI-Driven Innovation: Revolutionizing Furniture Design with FU Generator and Computational Ingenuity”

Niva Shrestha, Yige Liu(✉), and Philip F. Yuan(✉)

Tongji University, Shanghai, China
sniva2058@gmail.com, philipyuan007@tongji.edu.cn

Abstract. In contemporary design, the fusion of artificial intelligence (AI) and human creativity is generating unprecedented opportunities. This paper explores the application of AI in furniture design, specifically investigating how AI, inspired by natural elements, can revolutionize the creation of chairs. Drawing inspiration from the captivating aesthetics of waterfalls, this study weaves together digital intelligence and artistic vision to reshape traditional furniture design practices. The objective is to highlight the transformative potential of AI-driven innovation in furniture design, introducing new perspectives and inventive solutions. This exploration underscores the significant impact of technology on artistic endeavors, paving the way for novel approaches in the ever-evolving field of design.

Keywords: Computational design · AI image generation · Rattan chair · Waterfall inspiration · Innovation

1 Introduction

This research aims to investigate the convergence of technology and art in the context of furniture design, where the beauty of nature serves as a core inspiration for innovation. By focusing on the integration of artificial intelligence (AI) tools, specifically the FU Generator, this paper explores how AI can significantly enhance traditional design methodologies. Through this investigation, we seek to demonstrate the transformative potential of AI in introducing novel perspectives and dynamism into the design process, moving beyond conventional approaches. By leveraging the computational capabilities of AI alongside human creativity, this study presents a framework that redefines the boundaries of furniture design, offering new pathways for both aesthetic and functional innovation.

In our journey through the captivating world of furniture design, we’re not merely exploring chairs. Instead, we’re delving deep into the essence of design itself, pushing the limits of what we thought possible. We aim to demonstrate how artificial intelligence can breathe fresh energy into traditional methods, infusing our creative pursuits with vibrancy and igniting endless streams of inspiration (Fig. 9.1).

As we begin our exploration, we dive into how AI is changing the way we design, showing its ability to transform our creative process. Along the way, we discover the

J. Fu et al. (eds.), *AI for Architecture*, Computational Design and Robotic Fabrication,
https://doi.org/10.1007/978-981-95-0974-4_9

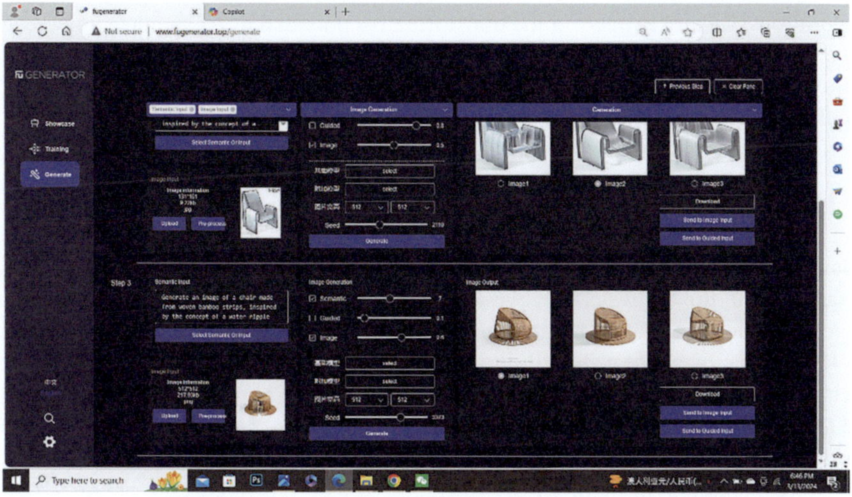

Fig. 9.1 FU generator image generation (*Source* Author)

many opportunities AI brings, reshaping how we think about design and opening up new possibilities.

2 Research Background

Furniture design has a rich history rooted in human creativity, drawing inspiration from diverse sources such as nature's beauty, cultural influences, and architectural marvels. Throughout the ages, designers have sculpted chairs, tables, and other furnishings, each piece a testament to their ingenuity and artistic vision.

In recent years, the advent of artificial intelligence (AI) has brought about a paradigm shift in various industries, including design. While AI has shown promise in fields like healthcare, finance, and transportation, its application in furniture design remains relatively uncharted territory.

Traditionally, furniture designers have relied on manual methods and their intuition to conceptualize and craft their creations. However, the integration of AI technologies opens up new possibilities, allowing designers to explore uncharted creative territories and optimize their designs for both form and function.

AI-driven tools, such as FU generator, Midjourney, Copilot, offer designers access to a vast array of design concepts and ideas. By inputting textual prompts or reference images, these algorithms can generate numerous design variations, providing designers with valuable inspiration and insights.

Despite the theoretical potential of AI in revolutionizing furniture design, practical implementations and real-world applications are still limited. While there have been discussions and explorations into the use of AI in design processes, there remains a gap between theoretical research and practical application.

This study seeks to address this gap by delving into the practical implications of integrating AI into the furniture design process. Through a series of experiments and explorations, we aim to uncover the transformative potential of AI in enhancing creativity, efficiency, and innovation in furniture design. By leveraging AI technologies, we aspire to push the boundaries of traditional design practices and pave the way for a new era of AI-driven creativity in furniture design.

3 Purpose

This journey begins with a vision: to capture the essence of nature in a chair. Inspired by the graceful flow of waterfalls, we set out to create something truly unique. But how do you translate the beauty of nature into a tangible design? That's where AI comes in.

By feeding images of waterfalls into the AI model, we unlock a world of possibilities. Each design is a testament to the power of digital creativity, blurring the lines between art and technology. Our goal is to create more than just a chair—we want to evoke emotion, to transport people to a place of wonder and awe.

As we embark on this journey, we're guided by a deep reverence for nature's beauty and a desire to harmonize with the natural world. With each iteration, we strive to capture the fluidity of water and the tranquility of cascading waterfalls. It's a journey of exploration and discovery, where every curve and contour tells a story of nature's boundless creativity.

Through the fusion of AI and human ingenuity, we're not just creating furniture—we're creating experiences. Our vision is to bring the beauty of nature into people's homes, enriching their lives and connecting them to the world around them. It's a bold vision, but one that we believe has the power to inspire, to uplift, and to transform the way we think about design.

4 Methodology and Experiments

In our quest to innovate furniture design, we embarked on a journey fueled by creativity and guided by rigorous experimentation. Our methodology embraced the fusion of AI technology, material science, and ergonomic principles to bring our vision to life. Here's a detailed account of our process:

4.1 Harnessing AI Creativity

At the heart of our methodology lies the innovative use of AI image generation, particularly through the FU Generator. We immersed ourselves in the captivating imagery of waterfalls, seeking inspiration for our chair design. By feeding simulated waterfall images into the AI model and providing textual prompts, we generated a diverse array of design concepts imbued with the dynamic spirit of cascading water. If initial outputs did not align with our vision, further refinements were made by modifying the prompts and adjusting parameters to yield more relevant designs. This navigation between variants, followed by manual adjustments in Rhino, ensured a final output that met our aesthetic and practical requirements (Fig. 9.2).

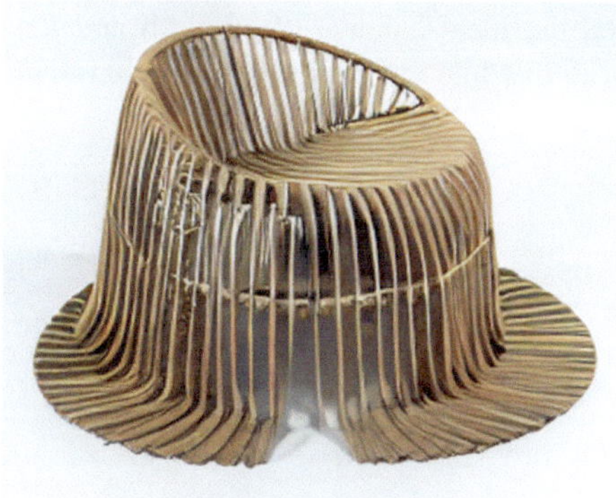

Fig. 9.2 Fu generator generated chair (*Source* Author)

4.2 Digital Sculpting with Rhino and Grasshopper

The 2D waterfall-inspired images generated by the FU Generator served as visual references. Translating these into 3D involved manually sculpting the forms in Rhino using Grasshopper to ensure that the organic nature of waterfalls was maintained while meeting ergonomic standards. Through iterative refinement, we balanced aesthetics with functionality, creating a design that seamlessly integrates digital creativity with real-world feasibility (Fig. 9.3).

Fig. 9.3 3D remodeling in Rhino (*Source* Author)

4.3 Exploring Material Properties

In the experimentation phase, we delved into the structural and aesthetic properties of both bamboo and rattan materials. While bamboo initially seemed promising due to its strength and flexibility, we encountered challenges in achieving the desired curvature inspired by waterfalls. Rattan, with its greater flexibility and pliability, emerged as the preferred choice for our design.

4.4 Optimization Through Trial and Error

Our experimentation with rattan involved exploring various dimensions and diameters to optimize the design. Through meticulous trial and error, we determined that a 2 cm diameter rattan provided the ideal balance of strength and flexibility, enabling us to achieve the desired curvature while maintaining structural integrity (Fig. 9.4).

Fig. 9.4 Optimization trials of rattan strips (*Source* Author)

4.5 Structural Analysis and Validation

Since the generated images from FU Generator did not inherently include mechanical considerations, it was essential to validate each design's structural integrity using Karamba. This ensured that the forms derived from artistic imagery were also structurally sound and could bear load-bearing conditions appropriately. Without this critical step, there would be a risk of translating arbitrary imagination into unsafe or impractical furniture. To validate the structural integrity of our chair design, we employed advanced simulation techniques using Karamba, a robust plugin integrated into Grasshopper. This pivotal step allowed us to subject our chair to various loading conditions, assessing its response to static and dynamic forces. Additionally, we explored the impact of different material dimensions on structural performance, adjusting parameters such as thickness and diameter to optimize strength and durability. Through iterative refinement, guided by simulation results, we ensured compliance with safety standards and performance expectations. Ultimately, this rigorous validation process has resulted in a chair design that seamlessly combines elegance, comfort, and reliability (Fig. 9.5).

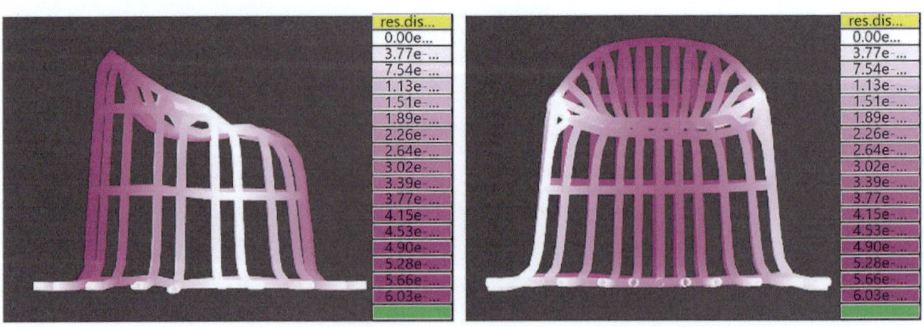

Fig. 9.5 Karamba simulation for structure analysis (*Source* Author)

4.6 Innovative Material Combination

Our construction approach involved strategically combining hard rattan in the bottom leg and soft rattan in the upper part of the chair. This decision was driven by principles of material mechanics, ergonomic design, and the need for easy bending. Hard rattan, with its denser and more rigid composition, provides essential load-bearing support, ensuring stability and durability. Soft rattan, besides offering flexibility and comfort for ergonomics, was chosen for its ease of bending, facilitating the intricate curves and bends required in the chair design. This combination optimizes structural integrity while prioritizing ergonomic comfort and ease of construction, aligning with both scientific and ergonomic principles (Fig. 9.6).

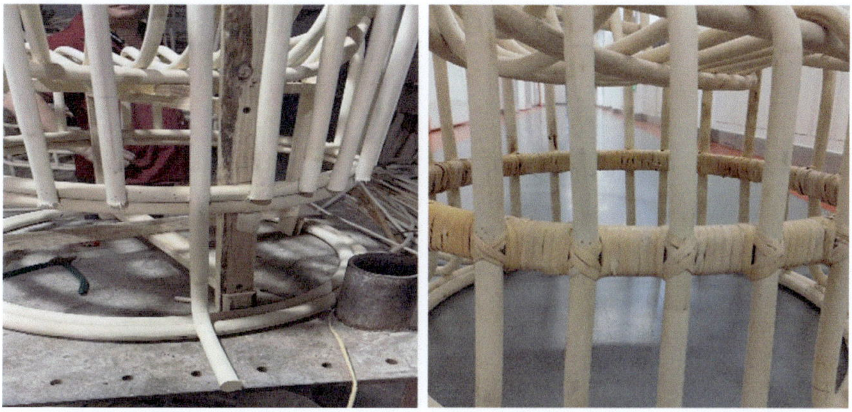

Fig. 9.6 Joining process of soft rattan and bottom hard rattan (*Source* Author)

4.7 Learning from Setbacks

Our journey was not without challenges. Initially, our reliance on soft bamboo for the entire chair design led to structural deficiencies, manifesting as a spring-like quality. However, this setback became an opportunity for innovation. By incorporating hard rattan for load-bearing support and soft bamboo for ergonomic comfort and easy bending, we addressed these concerns, resulting in a more robust and user-friendly chair design.

Through a blend of AI creativity, material exploration, and innovative construction techniques, we have redefined the possibilities of furniture design. Our chair stands as a testament to the power of interdisciplinary collaboration, paving the way for future innovations at the intersection of technology and craftsmanship (Fig. 9.7).

Fig. 9.7 Final chair design (*Source* Author)

4.8 Reproducibility of the Workflow

The workflow presented in this research can be reproduced using the same or similar tools and approaches. The initial waterfall imagery used as input for the FU Generator, followed by digital sculpting with Rhino, material testing, and structural validation with Karamba, offers a replicable process. By following these clearly defined steps, other designers can explore new themes and inspirations, potentially leading to a broader spectrum of AI-driven design outcomes (Fig. 9.8).

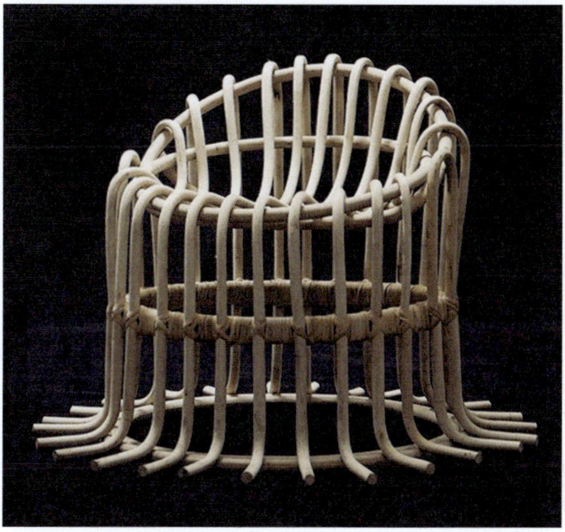

Fig. 9.8 Waterfall chair (*Source* Author)

5 Result

Our journey reaches its peak with a chair design that captures elegance and practicality. Inspired by the grace of waterfalls, our chair boasts symmetrical forms and a meticulous blend of materials, showcasing the beauty of nature in its simplicity.

By strategically utilizing hard rattan for load-bearing components and soft rattan for ergonomic comfort, we've achieved a perfect harmony between structure and usability. This thoughtful combination ensures not only durability and stability but also unparalleled comfort and support for the user.

Yet, our chair represents more than just a piece of furniture. It stands as a testament to our dedication to innovation and collaboration, pushing the boundaries of traditional design practices. It serves as a reminder of the power of creativity and the importance of embracing both technological advancements and our connection with the natural world.

6 Conclusion

Unlike more general models such as Stable Diffusion, which focus on broad image creation, FU Generator provides a more specialized tool tailored to the needs of designers in furniture and architecture. Its targeted training facilitates a smoother transition from creative conceptualization to practical realization, making it an invaluable asset in furniture design innovation.

In wrapping up our research, we've unveiled the incredible possibilities that AI can bring to furniture design. By tapping into AI's creative potential, we've opened doors to new ideas and ways of thinking about chairs. Our journey has shown that technology and nature can work hand in hand to inspire and innovate. Beyond just designing chairs, our exploration has deepened our appreciation for the connections between technology, nature, and human creativity. We've learned that there's still so much to discover and explore in this exciting field. As we look ahead, we're excited to continue pushing the boundaries of what's possible in design.

Acknowledgements. This research is funded by National Key R&D Program of China (2022YFE0141400) and National Key R&D Program of China (2023YFC3806900).

Bibliography

1. Smith J (2023) The impact of AI on furniture design. J Design Innov 5(2):45–56. DOI: 10
2. Brown A (2022) The role of nature in creative design. In: Designing with nature: exploring inspirations, pp 78–92. Design Press
3. OpenAI (2022) Introducing FU generator: AI for creative design. OpenAI Blog. https://blog.openai.com/fu-generator/
4. Rattan vs Bamboo Furniture: What Is the Difference? Furniture Maxi Blog. URL: https://www.furnituremaxi.com/blogs/news/rattan-vs-bamboo-furniture-what-is-the-difference
5. Karamba3D: Finite element analysis in grasshopper. URL: https://www.karamba3d.com/

Computational Urban Science

Research on Non-motor Vehicle Parking Around Subway Stations in Urban Suburbs Based on Microsimulation

Shuang Gao and Ziyu Tong(✉)

Nanjing University, Nanjing, China
1437502295@qq.com, tzy@nju.edu.cn

Abstract. As an important mode of transportation for the "last kilometer" of rail travel, due to the long distance between stations in urban suburbs, the demand for non-motor vehicle parking is increasing, and the unreasonable division of parking areas leads to the decline of traffic efficiency around subway stations. In this paper, the Dongda Chengxian subway station in suburban Nanjing is selected as the case study. Through field investigation and data collection, hybrid traffic simulation is conducted using Anylogic simulation software. Optimization strategies are proposed, and the proposed solutions are subjected to comparative analysis. The results show that by optimizing the parking of non-motor vehicles and increasing the effective road width, the pedestrian density can be reduced effectively and the travel time can be shortened by 7.35%. This study can provide reference for the planning and construction of non-motor vehicle parking areas around subway stations and improving the traffic efficiency of walking space.

Keywords: Suburban metro station surrounding · Non-motor vehicle parking · Mixed traffic simulation · Traffic efficiency · Scheme optimization

1 Introduction

In the context of travel in the era of energy conservation and emission reduction, the number of non-motor vehicles in cities is increasing, which brings convenience while the contradiction of planning and parking is also increasingly apparent. Under the influence of factors such as unclear demarcation of parking spaces, unbalanced demarcation of parking spaces and chaotic placement, the problem of non-motor vehicle parking around subway stations in suburban areas has become an important content that needs to be solved urgently in urban planning and construction.

The problem of non-motor vehicle parking has been the focus of many researches. This paper discusses the significance of regulating urban non-motor vehicle parking and the causes of urban non-motor vehicle parking [1]. Starting from the aspects of concept, facilities and management, the parking management strategy of electric bicycles around railway stations is systematically proposed [2]. In addition to the management level, the reasonable setting of non-motor vehicle parking area is the premise to solve the traffic congestion and improve the traffic efficiency, so it needs to be considered

J. Fu et al. (eds.), *AI for Architecture*, Computational Design and Robotic Fabrication,
https://doi.org/10.1007/978-981-95-0974-4_10

from the practical level, and the prerequisite for realizing this goal is to understand the riding behavior, walking behavior and parking behavior of non-motor vehicle users. Traffic simulation, as an important tool for evaluating and optimizing traffic systems, has attracted more and more attention from the academic community [3]. By studying non-motor vehicle individuals and their mixed flow characteristics, non-motor vehicle simulation models are established to rationally allocate non-motor lane resources and solve urban traffic problems [4]. By establishing the simulation model of mixed traffic flow in the parking section of single-vehicle road, the influence of non-motor vehicle flow and different parking forms on mixed traffic flow is analyzed [5]. By establishing a microscopic traffic flow model of on-road parking behavior, the impact of on-road parking on traffic flow and parking policies is simulated and studied, and management strategies and optimization measures for on-road parking are proposed [6]. Social force models of pedestrians, motor vehicles and bicycles are established to simulate mixed traffic and reflect different behavioral characteristics of various road users [7]. Bicycles mix with motor vehicles and pedestrians, affecting road traffic order and reducing road capacity, which may induce certain traffic accidents [8]. Therefore, it is urgent to improve the current situation of non-motor vehicle parking around subway stations in order to promote the traffic efficiency of walking space.

This study aims to propose a new idea to study the mixed traffic around metro stations in urban suburbs based on the social force model, combining cycling behavior, walking behavior and parking behavior to explore the impact of road and parking area Settings on the walking efficiency around metro stations, so as to provide a realistic basis for optimizing the safe traffic environment for non-motor vehicles to ride and walk. Promote sustainable urban transport development.

2 Case Study

2.1 Study Subjects

The problem of non-motor vehicle parking around subway stations in suburban areas is becoming increasingly obvious, and there has been a lack of targeted research. Therefore, the non-motor vehicle parking area around Dongda Chengxian subway station in suburban Nanjing is selected as the simulation space. There are 4 subway entrances and exits and an auxiliary elevator entrance and exit at this subway station. There are a large number of non-motor vehicle parking areas around the station, and there are a large number of non-motor vehicle cycling and walking phenomena, which will cause serious congestion problems.

The selection of this station has typical significance for this study.

2.2 Data Collection

Field research was conducted at the example site during the morning peak period of two consecutive weekdays from 7:30 to 8:30. In this time period, residents have the highest requirements for traffic efficiency. Firstly, the non-motor vehicle parking area and the maximum capacity of each parking area are measured on the spot. Secondly, through field observation and video processing technology, the behavior characteristics of non-motor vehicle cycling and walking in the natural state and the number of cyclists and walkers from all directions were recorded. Then, according to the field investigation statistics and literature reference, the age composition, walking speed, cycling speed, pedestrian size and non-motor vehicle size of the commuter population were determined, and the obtained data were used as parameters for mixed traffic simulation.

The Number of People Flow

There are 9 main starting points in the basic space, and the number of people flow is investigated at each starting point. The survey method adopted is the direct counting method, taking the number of people passing every hour as the basic unit, according to the number of people in the morning peak hours of working days in a week (Table 10.1).

Table 10.1 The number of people at each starting point during the working day commute

Ways	1	2	3	4	5	6	7	8	9
Walk	41	70	52	80	91	60	280	602	80
Bike	8	5	18	11	7	9	18	14	17
Electric vehicle	61	55	140	89	72	51	142	184	143
Total	110	130	210	180	170	120	440	800	240

Pedestrian Flow Composition

There is a significant difference in the age composition of subway passengers in the morning rush hour. The study on the age composition of commuters around the example stations has a crucial impact on the setting of walking and cycling speed parameters and the subsequent simulation modeling. According to the field survey and visit to the morning rush hour, in Dongda Chengxian subway station, 11.1% of people are under 20 years old, 53.8% are between 20 and 39 years old, 31.9% are between 40 and 59 years old, and 3.2% are over 60 years old. According to the survey statistics, the commuters in the example stations in the morning peak hours are mainly young and middle-aged people between 20 and 59 years old.

Pedestrian Flow Speed

The parameters of the non-motor vehicle social force model mainly include velocity and the related parameters of the interaction force and the boundary force. Under barrier-free conditions, the pedestrian's stride length is about 0.75 m. According to the data obtained from relevant research and field observation data, the expected speed of pedestrians in the non-motor vehicle area around subway stations in the morning rush hour is 0.7–1.7 m/s, and the expected speed interval of non-motor vehicles is set at 1.3–3.5 m/s.

Size of Pedestrian Flow

Most of the pedestrians in the morning rush hour of the railway station area are for commuting purposes. The survey found that there are almost no pedestrians carrying large luggage, only carrying backpacks or satchels, and the impact on pedestrian size is negligible. Therefore, the pedestrian size in this paper is based on pedestrian shoulder width as the simulation parameter. According to the Human Body Size of Chinese Adults (GB/T 10000–1998), the average shoulder width of Chinese adult males aged 18–60 is 375 mm, and that of adult females aged 18–55 is 351 mm. The size of non-motor vehicle will affect the avoidance distance, and then affect the interference degree of non-motor vehicle workshop, which is one of the factors that must be considered in the study of non-motor vehicle operation rule and the design of non-motor lane. According to the research and investigation, the average length of the electric bicycle body is 174 cm, and the average width of the body is 64 cm, which is significantly larger than the length of the conventional bicycle 170 cm and the width of 58 cm.

Other Parameters

By measuring the area of each non-motor vehicle parking area on the spot, combined with the size of non-motor vehicles, the maximum capacity of non- motor vehicle parking area at each station can be obtained. In addition, the subway station has an auxiliary elevator entrance, the running time from the ground floor to the negative first floor is 5s, and the average waiting time is 20s, which is also an indispensable important parameter in the simulation.

2.3 Establishment of Simulation Model

Mixed traffic flow simulation refers to the modeling of traffic scenarios involving multiple traffic participants. In this study, Anylogic is selected as the simulation platform. The combination of walking and cycling in the established simulation model integrates more complex individual behaviors and activities than the traditional traffic simulation model.

Construction of Basic Simulation Environment

Through the field measurement around Dongda Chengxian subway station, the environment model of non-motorized traffic system is established to provide a simulation system environment for the subsequent simulation. The elements involved in this step include obstacles, road network, parking area, starting and ending point of people flow, etc. (Fig. 10.1).

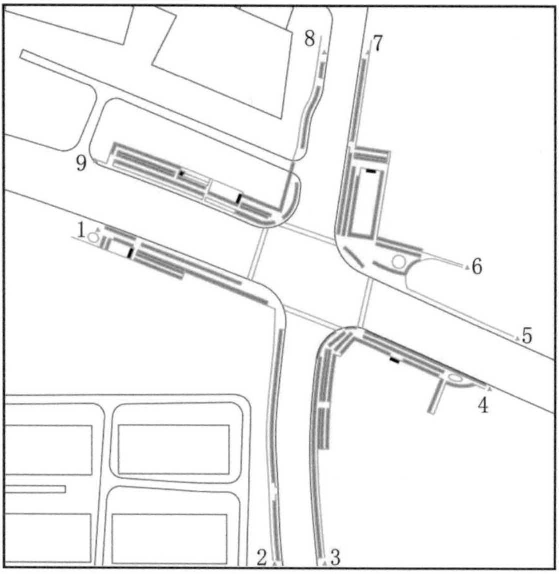

Fig. 10.1 Plan of Dongda Chengxian subway station

Behavioral Flow Modeling

In this study, the simulation of traffic behavior logic is the focus and difficulty of this study. Walkers walk directly from the initial point to the vanishing point, while non-motorized cyclists ride from the initial point to the non-motorized area, park the vehicle and then walk to the vanishing point. In the process of behavior modeling, it is necessary to test the passing behavior to ensure the authenticity and accuracy of the simulation effect, and to adjust and modify the situation in the model that is very different from the reality. The following is part of the behavior flow chart in this study (Fig. 10.2).

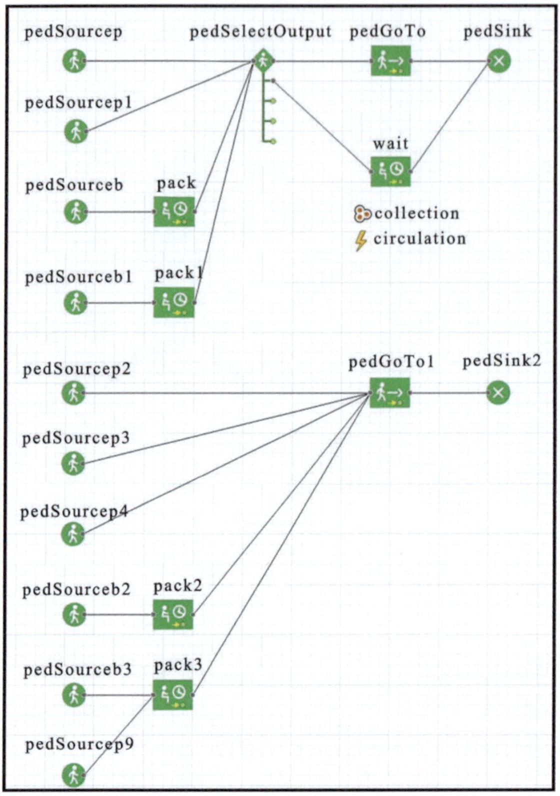

Fig. 10.2 Behavior flow chart

3 Simulation Results and Reason Analysis

In this study, two commonly used traffic indicators are selected, one is travel time and the other is crowd density.

Transit time refers to the time spent by passengers in the process of system appearance and disappearance, and its index can be obtained by Anylogic software. The shorter the passing time, the greater the traffic capacity of the section, indicating the better the passing efficiency.

Crowd density refers to the number of people per square meter in the passing process, and its index can be automatically generated by Anylogic software about the crowd density map in any time period in the study area.

A comparative analysis of the simulation diagram of the morning rush hour traffic system around Dongda Chengxian subway station shows that there is no significant change in the regional pedestrian density during the morning rush hour. The simulation 30 min is 8:00 a.m. in the morning rush hour, and the area with high pedestrian density is mainly concentrated in the subway entrance and parking area (Fig. 10.3). Simulation 60 min is 8:30 a.m. in the morning rush hour, and the crowd density at this moment is slightly higher than that at the simulation 30 min, and the crowd density in the parking area increases significantly (Fig. 10.4).

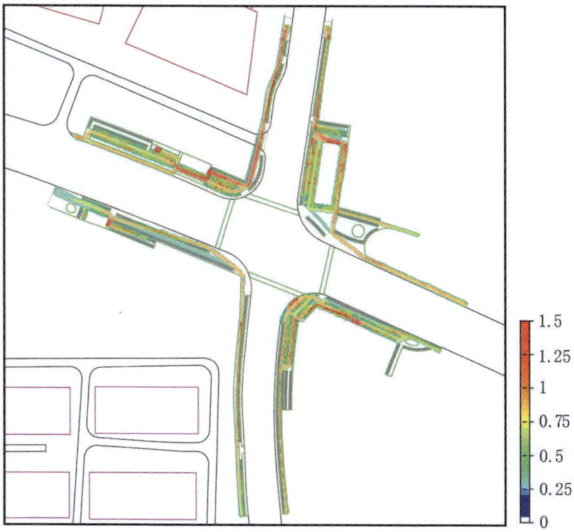

Fig. 10.3 Simulation of pedestrian density around the subway station after 30 minutes

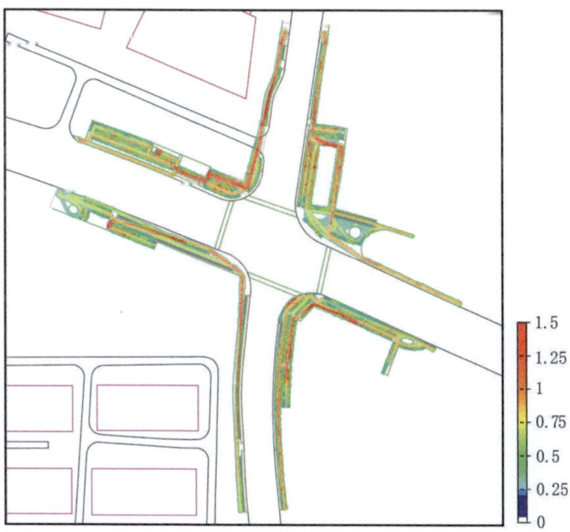

Fig. 10.4 Simulation of pedestrian density around the subway station after 60 minutes

Subway Exit 2: The simulation time of 30 minutes is 8:00 a.m. in the morning rush hour, and there is a density imbalance in the study range (Fig. 10.5). The simulation time of 60 minutes is 8:30 in the morning peak, and the regional pedestrian density increases significantly (Fig. 10.6). Concentrated congestion areas are corners and part of the road. The main reason is that there is a closed phenomenon at the end of the parking area, and unreasonable parking encroaches on the turning area, which is not easy to pass.

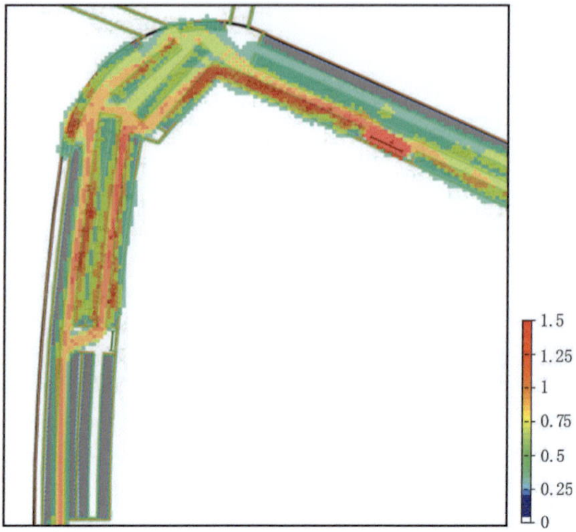

Fig. 10.5 Simulation of pedestrian density around the subway exit 2 after 30 minutes

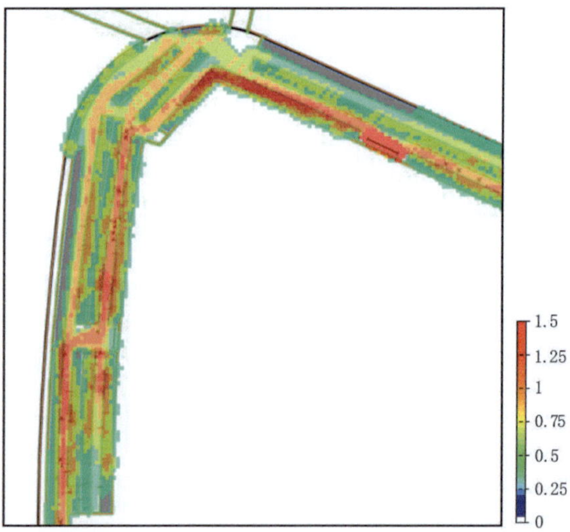

Fig. 10.6 Simulation of pedestrian density around the subway exit 2 after 60 minutes

Subway Exit 4: The simulated 60 min (Fig. 10.8) is 8:30 a.m. in the morning rush hour, and the pedestrian density in the area has no significant change compared with the simulated 30 min (Fig. 10.7). From the perspective of the overall population density, there is a large area of crowd crowding in the study area. The area of concentrated congestion area is the site entrance and exit and too narrow passing road. The main reason is that the effective width of some non-motorized lanes is too low, which is not conducive to the turnover of non-motor vehicles, thus affecting the traffic efficiency.

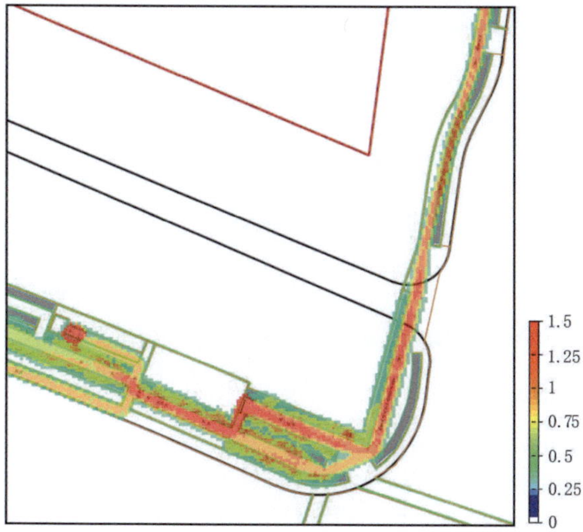

Fig. 10.7 Simulation of pedestrian density around the subway exit 4 after 30 minutes

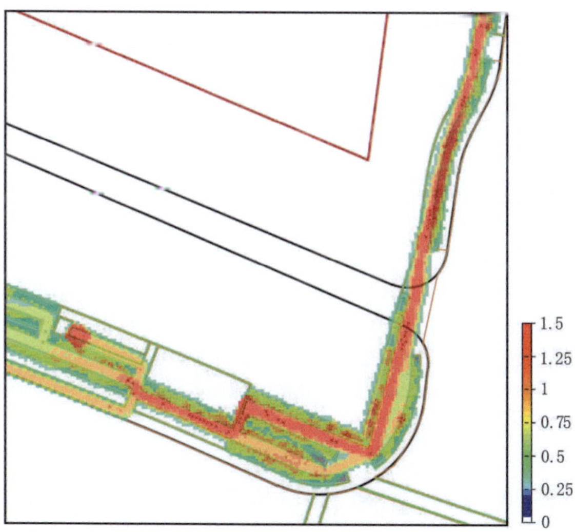

Fig. 10.8 Simulation of pedestrian density around the subway exit 4 after 60 minutes

Travel time for different modes of transportation at four exits are shown in the Table 10.2.

Table 10.2 Travel time for different modes of transportation at four exits

Average duration (s)	Exit 1	Exit 2	Exit 3	Exit 4
Cycling	118.67	125.83	98.4	120.23
Walking	158.12	106.93	82.68	124.04
Passing by	136.22	118.81	88.4	122.91

4 Optimization Strategy and Simulation Comparison

4.1 Optimization Strategy

Traffic is the basic function of traffic space, and the optimization of parking space should fully consider traffic efficiency. In the current situation, there is an unreasonable division of non-motor vehicle areas, which is not only conducive to the distribution of people, but to a large extent, it causes interference to the traffic flow line and intensifies the congestion near the entrance and exit. In the case of meeting the parking demand, the non-motor vehicle parking area of the station should be laid out according to the principle of traffic priority, and reasonable regional division should be carried out in the space to ensure that the needs of traffic, non-motor vehicle parking and pedestrian stay are met.

Reasonable Division of Parking Areas to Enhance Road Connectivity

The higher the connectivity of the road network system around the station, the stronger the accessibility of the walking space, and the higher the accessibility of the rail subway station. The end road or the broken road has little effect on the evacuation of pedestrian flow, and will affect the efficiency of pedestrian flow to a great extent during the period of continuous increase of pedestrian flow in the morning peak. Strengthen the communication of end-end roads or broken roads, play the role of road microcirculation, improve the connectivity of the road network, and improve the traffic efficiency of the traffic system. In the optimization plan, adjust the unreasonable parking Spaces, increase the connectivity of the road, and improve the parking efficiency and traffic efficiency of non-motor vehicles.

Moderately Increase the Pedestrian Space Scale of Subway Stations

On the one hand, the current "Urban Road Engineering Design Code" for the width of the sidewalk is mostly based on the grade of the road design, but there is no clear solution for the scale of the walking space around the subway station; On the other hand, in the preliminary design of urban planning, the influence of traffic facilities on the carrying capacity of the walking system is not fully considered, which results in the congestion and chaos of the walking system around the subway station. When the non-motor vehicle traffic volume is constant, the wider the non-motor vehicle lane, the smaller the non-motor vehicle traffic volume per unit width, and the lower the probability of non-motor vehicles occupying the road. In the optimization scheme, the non-motorized lane can be expanded to ensure the effective width of the non-motorized lane.

4.2 Simulation and Comparative Inspection of the Improved Scheme

Scheme 1: According to the above optimization and improvement measures for Dongda Chengxian Subway Station, simulation modeling is carried out on the optimized non-motor vehicle parking space to test the results of the optimization scheme (Fig. 10.9)

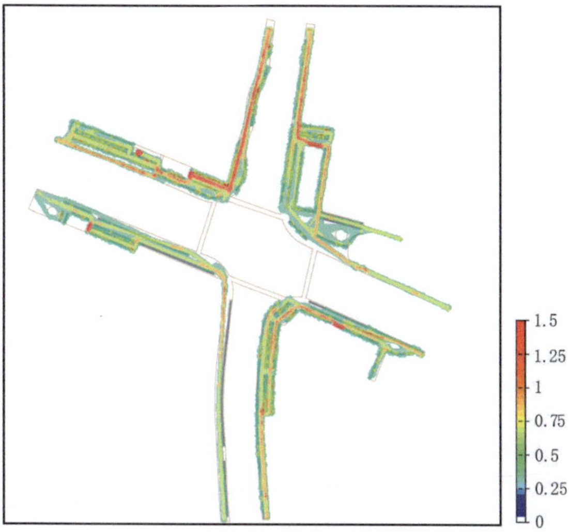

Fig. 10.9 Simulation of pedestrian density around the subway station after 60 minutes following optimization

The optimization plan is simulated and tested. The pedestrian density map is output at the peak of the pedestrian density. Compared with the previous pedestrian density map, it can be found that the original location with excessive pedestrian density in the region has been significantly alleviated, and the traffic efficiency has been greatly improved.

Compared with the traffic time before optimization derived by anylogic, the traffic time of both cycling and walking is significantly shortened. The total average travel time is shortened by 7.96s, in which the average walking time is shortened by 7.34s and the average cycling time is shortened by 8.8s (Table 10.3).

Table 10.3 Travel time for different modes of transportation at four exits following optimization

Average duration (s)	Exit 1	Exit 2	Exit 3	Exit 4
Cycling	112.41	111.45	92.32	112.9
Walking	124.23	103.78	79.26	115.75
Passing by	118.29	108.33	83.97	114.8

Scheme 2: Optimize again on the basis of plan 1, and adjust some crowded parking areas while ensuring a constant maximum number of parking. The results show that the average time of total passage is shortened by 0.79 s

In this study, we simulated three scenarios. One scenario is the real status quo, and the other two scenarios are the optimized scenarios. Before the simulation, we have measured various data around Dongda Chengxian subway station on site, including the width of different lanes, the size and maximum parking capacity of parking areas, the number of pedestrians and cyclists in each exit direction during the morning rush hour on working days, and set parameters such as cycling, walking speed, and waiting time for the auxiliary elevator entrance at Exit 4. Finally, both of these conditions were simulated.

The results show that the comparison before and after optimization reflects the improvement of traffic efficiency under the guidance of the optimization strategy in this paper to a large extent. As can be seen from the density map, the original location of excessive traffic density has been significantly alleviated. According to the above statistical data of average passing time before and after optimization, the average passing time after optimization is reduced by 7.35%.

5 Conclusion

The main findings of this study are summarized as follows: reasonable arrangement of non-motor vehicle parking spaces and appropriate increase of effective road width can significantly reduce traffic density and save travel time.

The new method proposed in this paper combines real-world data with agent- based modeling technology to realize advanced hybrid simulation of non-motor vehicle riding and walking behavior, including non-motor vehicle riders' behavior of walking after parking their vehicles, which can more clearly realize the information exchange between different individuals and the interaction between individuals and the simulated environment. This can provide a methodological reference for the future study of the traffic efficiency of the pedestrian system. Because the periphery of suburban subway stations undertakes the commuting needs of pedestrians to the urban area during the morning rush hour, and there are common problems such as crowding and low traffic efficiency around the station, this method is also suitable for solving the traffic efficiency problems around suburban subway stations in other cities, which provides new thinking and inspiration for the future spatial planning and design of parking around similar stations in cities. In this way, planners can dynamically simulate the traffic conditions in the non-motorized parking area around the site, so as to rationalize the transportation infrastructure, and also identify and optimize the current transportation system.

However, there are still some disadvantages. The parameter setting in the simulation model requires more experience to make the simulation results more objective and accurate. In addition to this, the modeling did not take into account the small number of people leaving the suburban subway station during the morning rush hour. Through further research on these directions, we can provide effective theoretical support and technical approaches for the improvement of urban non-motor vehicle parking.

References

1. Zehui Q, Yuyun H, Shuhua M (2023) Research on status quo, problems and Countermeasures of urban non-motor vehicle parking management [J]. Indus Technol Forum 22(19):183–184 (in Chinese)

2. Dong W, Lei Z, Wang L (2022) Management strategy of electric bicycle parking around railway stations in Shenzhen [C]. Urban transportation planning academic committee of urban planning society of China. Green, wisdom, fusion—2021/2022 Chinese urban traffic planning conference proceedings. [publisher unknown], 2022, 11. https://doi.org/10.26914/Arthur c. nkihy. 2022.059258
3. Bazghandi A (2012) Techniques, advantages and problems of agent based modeling for traffic simulation [J]. In J Comput Sci Issues (IJCSI) 9(1):115–119
4. Qiaobing Y (2020). Research on Non-motor vehicle Mixed flow Simulation based on Social force Model [D]. Building of Beijing university. 10.26943/, dc nki. GBJZC. 2020.000472.5
5. Jiarong W (2018) Research on simulation modeling of the impact of on-road parking behavior on road traffic flow [D]. Beijing Jiaotong University
6. Xiaofei Y (2011) Research on the impact of on-road parking on road traffic and its management strategy [D]. Nanjing Forestry University
7. Qianqian Y (2019) Human-vehicle hybrid traffic simulation based on social forces [D]. Zhejiang University
8. Klassen J, El-Basyouny K, Islam TM (2014) Analyzing the severity of bicycle-motor vehicle collision using spatial mixed logit models: a City of Edmonton case study [J]. Safety Sci, 62295–304

Solar Energy Potential Prediction Model of Array Office Block Based on Ensemble Learning Algorithm

Gaomei Li[1,2,3,4], Qiuguo He[3], and Shen Xu[3,4(✉)]

[1] School of Architecture, Tsinghua University, Beijing, China
[2] Key Laboratory of Eco Planning & Green Building, Ministry of Education (Tsinghua University), Beijing, China
[3] School of Architecture and Urban Planning, Huazhong University of Science and Technology, Wuhan, China
46906543@qq.com
[4] Hubei Engineering and Technology Research Centre of Urbanisation, Wuhan, China

Abstract. In the context of the artificial intelligence era, machine learning algorithms can be used to achieve rapid prediction of the solar potential of urban blocks. However, most of the existing studies used computer simulation, this paper aims to construct a prediction model of the solar potential of urban blocks based on machine learning algorithms. This paper constructs a parametric model of office blocks based on real block cases in Wuhan city, and develops an arrayed solar potential prediction model for office blocks based on ensemble learning algorithms. The results showed that the ensemble learning prediction model can achieve rapid prediction of solar potential, achieve rapid feedback of solar potential of the design scheme, and significantly improve the efficiency of the performance evaluation and design optimisation of the urban block design scheme; the R^2 of the solar potential prediction model for the roof and façade are 0.99 and 0.98, respectively, and the prediction model has a high degree of accuracy. This study aims to provide a decision-making basis for the planning and design of urban office blocks, improve the scientific and rational design, and help urban office blocks achieve carbon neutrality.

Keywords: Office block · Machine learning algorithm · Ensemble learning · Solar potential · Block morphology · Prediction model

1 Introduction

In response to global climate change and the energy crisis, in September 2020, President Xi Jinping made a solemn pledge at the 75th session of the United Nations General Assembly that China's carbon dioxide emissions would strive to peak by 2030 and work towards carbon neutrality by 2060 [1]. The 2023 China Building Energy Efficiency Annual Development Research Report [2] stated that the total commodity energy consumption of building operation in 2021 was 1.11 billion tec, accounting for about 21%

J. Fu et al. (eds.), *AI for Architecture*, Computational Design and Robotic Fabrication,
https://doi.org/10.1007/978-981-95-0974-4_11

of the national energy consumption, with a total carbon emission of 2.2 billion tCO_2, and the average carbon emission target of building operation per unit area was 32 kg CO_2/m^2. In the context of the 30–60 dual-carbon strategy goal, the adoption of distributed photovoltaics, a clean energy source, is a key link in realising the energy transition [3, 4]. Building photovoltaic integration can realize the self-production and self-use of photovoltaic power generation, and provide energy for block buildings. Achieving carbon neutrality in buildings requires an update of the whole process of building design, which needs to focus not only on 'robustness, practicality and aesthetics', but also on the green building performance of the design scheme [5]. The traditional performance evaluation of building solutions is based on the process of 'design—simulate—optimise design-re-simulate', which is a time-consuming and labour-intensive performance evaluation process using computer simulation methods. Currently, the assessment of solar energy potential of urban blocks in the pre-design stage is generally carried out by computer simulation [6, 7], which is difficult to achieve real-time prediction of the design scheme, and there is a time lag in guiding the architects to modify the design scheme. Building performance prediction models based on machine learning algorithms can achieve rapid prediction of the performance of building schemes, realising the 'drawing and simulation' of building schemes and improving the efficiency of architects' design decisions [8, 9].

Therefore, in the context of the construction of low-carbon urban areas and the continuous development of solar buildings, this paper is based on the architect's programme design needs in the pre-design stage, takes the arrayed (pavilion-type and slab-type) office blocks in Wuhan as the study object, constructs the parametric model of the office blocks, taking into account the urban meteorological conditions, the performance of photovoltaic modules, and the block's multi-scale design parameters, and puts forward a machine-learning algorithm-based rapid prediction model for the roof and façade solar potential of office blocks. The aim of this paper is to enable rapid prediction of the solar potential of building schemes in office blocks and to provide architects with design aids to improve decision-making efficiency.

2 Methodology

This section proposes a process for constructing a solar potential prediction model for arrayed office blocks based on an ensemble learning algorithm, as shown in Fig. 11.1. Step 1, 96 arrayed office blocks in Wuhan are classified based on the building typology method, and typical group units are extracted with ideal block site construction, relying on the Grasshopper platform, and the Python programming language is used to generate the parametric model of the office blocks; Step 2, a workflow for solar potential assessment of urban office blocks is constructed based on the Rhino&Grasshopper parametric platform; Step 3, the multi-scale design parameters are extracted and calculated of the arrayed office blocks, including the three scales of block- building-facade; Step 4, the design parameters of the office block are integrated with the solar potential to construct the training and test sets of the prediction model; Step 5: The solar potential prediction model for the office block is developed based on ensemble learning algorithms using the office block multi-scale design parameters, climate parameters, and PV module performance as inputs and the roof and façade solar potential as outputs.

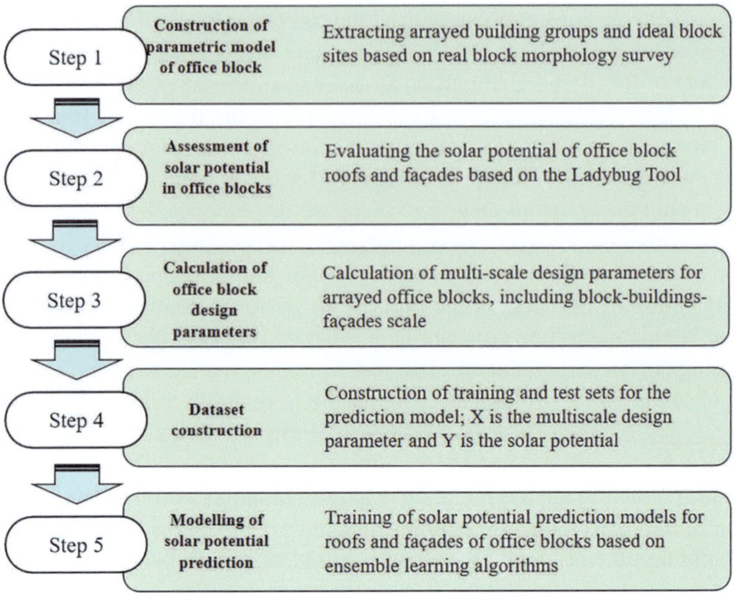

Fig. 11.1 Workflow of an arrayed office blocks solar potential prediction model constructing based on ensemble learning algorithms

2.1 Construction of Parametric Model of Office Blocks

Extraction of Typical Forms of Building Groups

Based on the field investigation of the real office block form in Wuhan city, the typical building group form is extracted, and the parametric model of office block is built. Based on literature research [10, 11], it is known that office block forms can be classified into pavilion, slab and enclosed according to the building typology approach. In this paper, the pavilion and slab blocks, which account for the larger part of the study, is taken as the study object. A combination of satellite maps, field research, street maps and open street maps was used to analyse 96 pavilion and slab blocks in Wuhan, extracting information on design parameters such as building group layout, building width, depth, number of floors, height of floors, etc., and determining the range of intervals for the design parameters of pavilion and slab blocks. Pavilion and slab blocks have consistency in parametric generation logic and are therefore collectively referred to as arrays. The typical examples of arrayed office blocks are shown in Fig. 11.2.

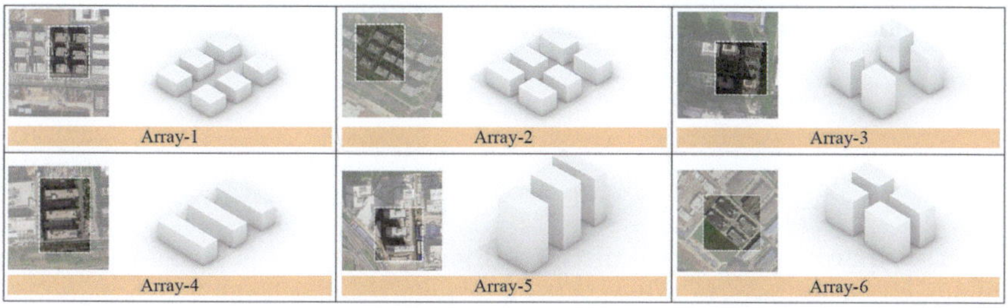

Fig. 11.2 Typical examples of arrayed office groups

Construction of Parametric Model of Office Blocks

The assessment of the solar potential of the office block needs to take into account not only the target building, but also the shading of the target building by the surrounding buildings, so the study area and the surrounding environment need to be considered when constructing the parametric model of the office blocks. The survey results of the real block cases indicated that the site size scale of the office building group is 90–150 m, combined with the diversity of morphological changes of the building group, this paper sets the site of the office building group as 145 m. The four building groups form the ideal office block, which is surrounded by the urban feeder roads, and by the surrounding environmental groups that are consistent with the study area. The schematic of the parametric model of an arrayed office block is shown in Fig. 11.3.

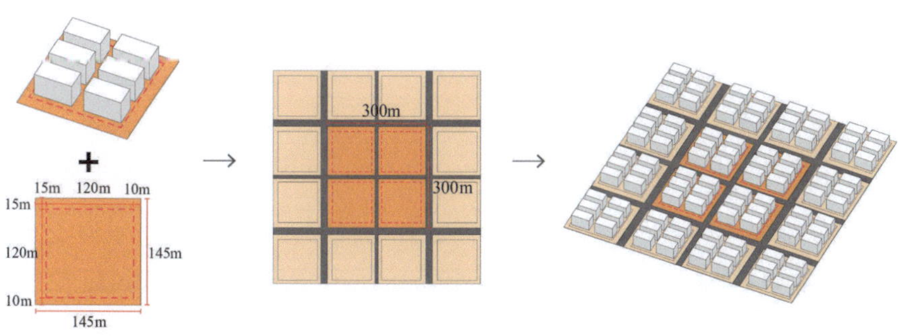

Fig. 11.3 Schematic of the parametric model of an arrayed office block

According to the architect's design logic, the parametric model generation of the office block can be divided into the following steps: firstly, according to the design assignment and design specifications to determine the distance of the building back from the urban road, to determine the scope of the building design; Secondly, the building units are arranged within the design scope, and the number of buildings is determined based on the standard floor area; finally, there is building accounting, i.e., the north–south and east–west spacing of the buildings should be in accordance with the fire code, and when they are not in accordance, the buildings are automatically reduced by one row or one column.

The building arrangement of an arrayed office block satisfies the following equation:

$$a + 2(d - NS) + m * BD + (m - 1)BI(NS) + b \leq L_1 \tag{11.1}$$

$$a + 2(d - EW) + n * BW + (\text{n - 1}) + b \leq L_2 \tag{11.2}$$

where, a and b are building setback distances from urban roads; d-NS and d-EW represent building setback distances from the red line; BD is the building depth; BW is the building width; BI(NS) and BI(EW) denote the building north–south distance and the building east–west distance; m and n refer to the number of rows and columns of the building. The following conditions need to be met when carrying out building accounting, when the building height is greater than or equal to 24 m, the building spacing is not less than 13 m; when the building height is less than 24 m, the building spacing is greater than or equal to 6 m and less than 13 m (Fig. 11.4).

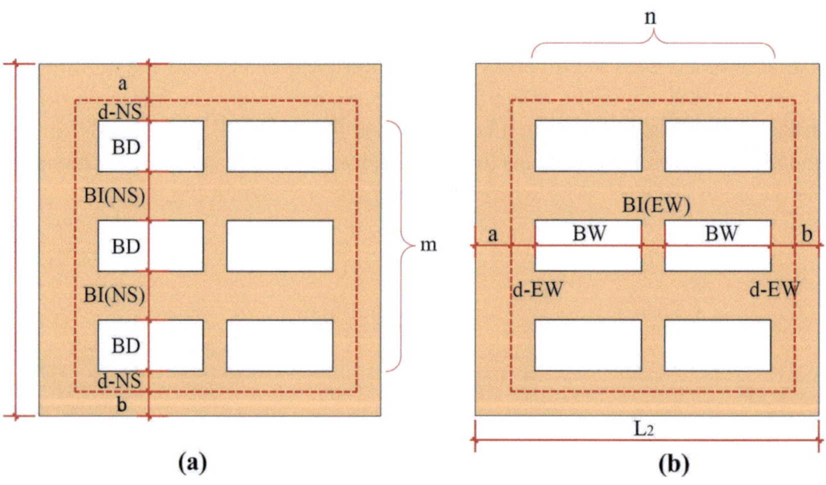

Fig. 11.4 Parametric model generation constraints for arrayed office blocks

2.2 Assessment of Solar Potential in Office Blocks

Solar potential refers to the process by which photovoltaic (PV) panels mounted on the surface of a building convert solar radiation into electrical energy. In this paper, based on the results related to the assessment of solar potential in blocks of our research group [6, 12], the solar potential of urban blocks is classified into three levels: radiation potential, installation potential and technical potential. Radiation potential refers to the effective solar radiation received by the building surface, which is related to the radiation threshold, which is the minimum radiation value of PV panels mounted on the building surface to satisfy the input–output balance of the PV system over the whole life cycle. Solar radiation can be achieved by a block solar assessment tool based on the Grasshopper parameterisation platform employing Ladybug Tools, which has been shown to be highly accurate by Shen Xu [13]. Installation potential refers to the area of the building surface suitable for mounting PV modules, the indicator that is closely related to the mountable area of the building surface in the block as well as to the radiation thresholds, with the installation factor obtained through a survey of real block cases. The technical potential represents the annual energy generation of a PV module mounted on the surface of a block building, which is influenced by a combination of radiation potential, installation potential, photovoltaic module photovoltaic conversion rate, overall efficiency, attenuation rate and life cycle. In this paper, the monocrystalline silicon PV module with the highest market share is selected for analysis, the life cycle of the PV module is set to 25 years, and the radiation threshold is 568 kWh/m^2/y. The parameter settings for the solar radiation simulation are shown in Table 11.1.

Table 11.1 Parameter settings for solar radiation simulation

Relevant parameters	Parameter settings
Meteorological parameters	CHN_Hubei.Wuhan.574940_CSWD
Simulation period	1月1日 00:00 至 1 2 月 31 日 24:00
Grid size	1 m * 1 m
Building surface reflectance	0.2
Ground reflectance	0.2

Office block solar potential refers to the technical potential and is calculated using the following equation:

$$E_{\mathrm{p}} = H_A \times A_{pv} \times \eta_{sys} \times K \times \sum_{1}^{N} (1 - R_1 - (N-1) \times R_d) \tag{11.3}$$

where, E_p is technical potential, H_A represents the cumulative solar radiation received by the building surface, A_{pv} refers to the mountable area of the PV modules, η_{sys} denotes the photovoltaic conversion rate of the PV modules, which is taken as 21.86 per cent, K refers to the combined efficiency factor, which takes the value of 85 per cent, R_1 and R_d mean the attenuation rate of the PV module in the first year and after, taking the values of 1.5% and 0.48%. In order to make the PV power generation correspond to the building energy consumption, the solar potential (technical potential) in this paper refers to the PV power generation per unit of building area.

2.3 Calculation of Office Block Design Parameters

In order to correspond to the urban block design process of 'sketch design—general layout design—preliminary design of building unit scheme—perfect design of building unit scheme', this paper selects the design parameters related to the design process, describing the intensity of construction, site layout, plan morphology, building height, building form and façade design, including the three scales of the block—building unit—building skin. -building skin at three scales, including building cover ratio (BCR), floor area ratio (FAR), building shape factor (BSF), open space ratio (OSR), block orientation (O), roof to façade area ratio (R/F), building base area (BA), building north–south interval

(BI(NS)), building east–west interval (BI(EW)), standard floor area (Sa), building depth (BD), building number of floors (FNa), building height (FH), window-to-wall ratio in each direction (WWR-N, WWR-S, WWR-E, WWR-W). Of these, the open space ratio is defined as the ratio of the open space area of an office block to the total floor area.

2.4 Office Blocks Dataset Construction

The office block dataset consists of two parts: the design parameter dataset and the solar potential dataset. The parametric model of the office block is generated from the controlling parameters, and then the parametric model is simulated to obtain the solar potential, including the roof solar potential and the façade solar potential. The design parameters include controlling parameters and descriptive parameters. Among the 17 design parameters, block orientation, building north–south interval, building east–west interval, standard floor area, building depth, building number of floors, building height, and window-to-wall ratio in each direction are controlling parameters, and the rest are descriptive parameters. Through the survey and analysis of the real case of array office block in Wuhan, the threshold range and step change of each controlling parameter are determined, and 500 groups of controlling parameter data are obtained by Latin hypercubic sampling, which are inputted into the parametric model to generate the corresponding 500 parametric models, and the simulation of the solar radiation is carried out based on the Grasshopper parametric platform to obtain the solar energy potential data of the roofs and façades. In addition to this, descriptive parameters are calculated for each of the 500 office block parametric models, and the descriptive parameters are integrated with the control parameters to generate the design parameter dataset. The design parameter dataset and the solar potential dataset are integrated to construct the office block dataset. The threshold ranges and step changes for the controlling parameters are shown in Table 11.2.

Table 11.2 Threshold ranges and step changes in controlling parameters for arrayed office blocks

Parameters	Threshold range	Step	Parameters	Threshold range	Step
O	−90° –90°	5°	FH	3.8~5.4 m	0.1 m
BI(NS)	7–52 m	1 m	WWR-E	0.2–0.9	0.1
BI(EW)	6–21 m	1 m	WWR-S	0.2–0.9	0.1
Sa	400–2500 m^2	100 m^2	WWR-W	0.2–0.9	0.1
BD	14–34 m	1 m	WWR-N	0.2–0.9	0.1
FNa	3–25	1			

2.5 Construction of Solar Energy Potential Prediction Model for Office Blocks

Ensemble learning algorithms imply the use of multiple base learners for the learning task, which has the advantages of high accuracy, low risk of overfitting and good robustness. In this paper, six ensemble learning algorithms were used to construct the solar potential prediction model for arrayed office blocks, including Bagging algorithm, Extremely randomized trees algorithm, Random forest algorithm, Gradient Boosting algorithm, AdaBoosting algorithm and XGBoosting algorithm. The bagging algorithm is an adaptive technique that generates multiple subsample sets by introducing bootstrap sampling method and constructs a base model based on each subsample set. Its key advantage is that it reduces model overfitting. By integrating multiple base models, the Bagging algorithm is better able to capture the complexity and uncertainty of the data, thus reducing the risk of model overfitting. The Extremely randomized trees algorithm is an ensemble learning method that improves the generalisation and stability of a model by constructing multiple extremely randomised decision trees. The Random forest algorithm is an ensemble learning method based on decision trees that improves the performance of a model by combining multiple decision trees together. The core principle of Random forest is to reduce overfitting by generating multiple independent decision trees and averaging them over the training data. The Gradient Boosting algorithm is an ensemble learning method based on convex optimisation that improves the performance of a model by combining multiple weak learners. The core principle of Gradient Boosting is to progressively optimise the model by performing gradient descent on each weak learner. AdaBoosting algorithm is the process of iterating weak classifiers into strong classifiers with full consideration of the weights of each classifier, changing the sample weights and classifier weights during each iteration, and the final result is the sum of the weights of each classifier. The XGBoosting algorithm is an optimised distributed gradient boosting library which has the advantage of being efficient, flexible and portable.

The dataset was divided into a training set and a test set in the ratio of 70% and 30%. The dataset was fed into six ensemble learning algorithms for training, evaluated using R^2 and MAE metrics, and the algorithms with good evaluation metrics were selected to construct the solar potential prediction model for office blocks.

3 Results

3.1 Sample Dataset Characteristics

Parametric model construction and solar radiation simulation (both roof and façade) were carried out on 500 sets of data sampled from Latin Hypercube, and after excluding the anomalous results, there were 465 valid data, which were divided into training and test sets in the ratio of 70% and 30%. The samples were used as input to the ensemble learning algorithms to train the solar potential prediction model.

Figures 11.5 and 11.6 depicts the distribution characteristics of representative design parameters (building cover ratio, floor area ratio, building shape factor, standard floor area, number of floors, and south-facing window-to-wall ratio) for the training and test sets, and the results indicated that the design parameters are evenly distributed, and it proves the reasonableness and stochasticity of the dataset division.

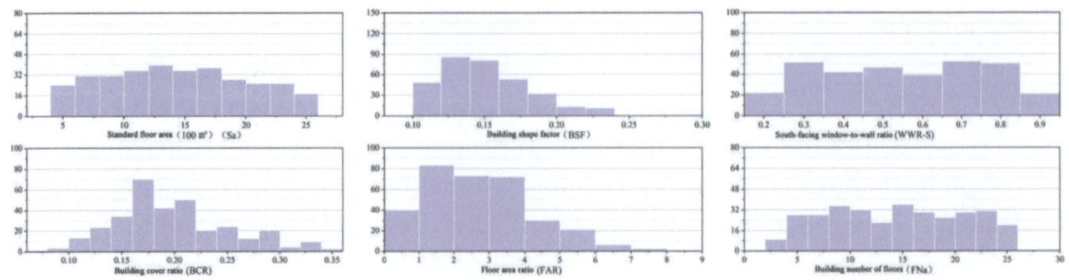

Fig. 11.5 Distributional characteristics of representative design parameters for the training sets

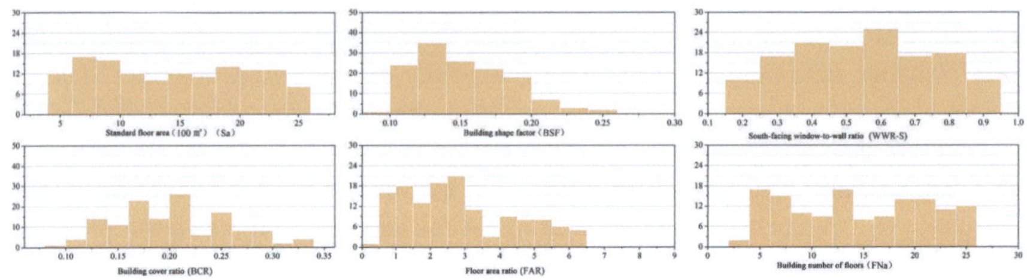

Fig. 11.6 Distributional characteristics of representative design parameters for the test sets

3.2 Sample Dataset Analysis

The 17 arrayed office blocks multi-scale design parameters were correlated with the roof solar potential and façade solar potential respectively, and the results of the Pearson correlation analysis are shown in Table 11.3.

Table 11.3 Correlation analysis of influencing factors

Morphological parameters	EGI-R(r)	Sig	Morphological parameters	EGI-F(r)	Sig
BCR	0	0.992	BCR	−0.343**	0.000
FAR	−0.721**	0.000	FAR	−0.608**	0.000
BSF	0.475**	0.000	BSF	0.369**	0.000
OSR	0.827**	0.000	OSR	0.629**	0.000
R/F	0.925**	0.000	R/F	0.472**	0.000
BA	0	0.992	BA	−0.343**	0.000
O	0.086	0.063	O	−0.151**	0.001
BI(NS)	0.026	0.581	BI(NS)	0.237**	0.000
BI(EW)	0.031	0.506	BI(EW)	0.093*	0.045
Sa	−0.038	0.415	Sa	−0.102*	0.028
BD	0.033	0.479	BD	−0.084	0.072
FNa	−0.857**	0.000	FNa	−0.549**	0.000
FH	−0.043	0.355	FH	0.009	0.849

(*continued*)

Table 11.3 (*continued*)

Morphological parameters	EGI-R(r)	Sig	Morphological parameters	EGI-F(r)	Sig
WWR-N	0.009	0.847	WWR-N	0.004	0.938
WWR-W	−0.016	0.732	WWR-W	−0.289**	0.000
WWR-S	0.051	0.271	WWR-S	−0.323**	0.000
WWR-E	0.008	0.872	WWR-E	0.052	0.265

The design parameter that has the highest correlation with roof solar potential is the roof to façade area ratio (R/F), which is positively correlated, this is because when the roof to façade area ratio (R/F) is larger, it means that the roof area is larger, i.e., the roof solar potential is larger. This was followed by the building number of floors, which is negatively correlated, then open space ratio (positively correlated), floor area ratio (negatively correlated), and building shape factor (positively correlated). For façade solar potential, the design parameter with the highest correlation is the open space ratio (0.629), this is because when the open space ratio is higher, it means that the building spacing is higher, the mutual shading between buildings is lower and the façade solar potential is higher. This was closely followed by floor area ratio (−0.608), building number of floors (−0.549), roof to façade area ratio (0.472), building shape factor (0.369), building cover ratio (−0.343), building base area (−0.343), south-facing window-to-wall ratio (WWR-S), and west-facing window-to-wall ratio (−0.289). When these design parameters are positively correlated with solar potential, the solar potential increases as these parameters become larger, and conversely, when negatively correlated with solar potential, the solar potential decreases as these parameters become larger.

3.3 Arrayed Office Block Ensemble Learning Solar Potential Prediction Models

The dataset was normalised in order to eliminate the effect of differences in the magnitude of the design parameters on the solar potential. The processed dataset was divided into training set and test set according to the ratio of 70% and 30%, the training set was fed into the debugged ensemble learning algorithm for training, the goodness of fit (R^2) and mean square error (MSE) were used as the evaluation indexes, and the trained model was tested for generalisation with the test set, the algorithmic model that is optimal in terms of R^2 and MSE in both the training set and the test set was selected for the solar energy potential prediction. The R^2 and MSE indicators for the six ensemble learning algorithms are shown in Tables 11.4 and 11.5, and the highest rated algorithms (Random Forest, Gradient Boosting) were selected as the predictive models for the solar potential of the roof and façade.

Table 11.4 Comparative analysis of the scoring results of roof solar potential prediction models

Algorithm types	R^2		MSE	
	Training sets	Test sets	Training sets	Test sets
Bagging algorithm	0.99	0.99	0.01	0.04
Extremely randomized trees algorithm	1.0	0.99	7.57	0.04
Random forest algorithm	0.99	0.99	0.01	0.03
Gradient Boosting algorithm	0.99	0.99	0.01	0.04
AdaBoosting algorithm	0.99	0.99	0.21	0.18
XGBoosting algorithm	0.99	0.99	7.05	0.04

Table 11.5 Comparative analysis of the scoring results of façade solar potential prediction models

Algorithm types	R^2		MSE	
	Training sets	Test sets	Training sets	Test sets
Bagging algorithm	0.96	0.66	0.16	0.91
Extremely randomized trees algorithm	1.0	0.71	2.12	0.76
Random forest algorithm	0.97	0.70	0.13	0.79
Gradient Boosting algorithm	0.98	0.74	0.08	0.68
AdaBoosting algorithm	0.84	0.55	0.64	1.19
XGBoosting algorithm	0.99	0.70	8.11	0.79

4 Conclusion

This paper presents the ensemble learning algorithm model for predicting the solar potential (roof and façade) of an array office block. The study considers the meteorological conditions of the city, PV module types and multi-scale design parameters for the prediction of PV power generation per unit of floor area. The R^2 and MSE of six ensemble learning algorithms were compared and analysed, and the highest rated algorithmic model was selected for the prediction of solar energy potential in office blocks. Compared with the existing linear prediction model for solar potential constructed based on block morphology parameters for campus dormitory blocks [14] ($R^2 = 0.73$), the machine learning prediction model constructed in this paper has higher precision and accuracy, which may be attributed to the non-linear characteristics of the machine learning algorithm compared to the linear model. Compared with the ensemble learning algorithm prediction model for different functional blocks in urban Shanghai [15] (XGBoosting algorithm, $R^2 = 0.91$), the prediction model constructed in this paper has a better accuracy, which may be due to the fact that the block morphology parameters selected in this paper are more reflective of the characteristics of solar energy potential.

In the future, the model can be combined with multi-objective optimisation to achieve solar potential performance-driven office block design scheme generation. This study aims to provide useful insights and references for architects to carry out solar planning and design of office blocks and the construction of zero-carbon blocks.

Acknowledgements. This research was supported by the National Natural Science Foundation (No. 52378020); the Fundamental Research Funds for the Central Universities, China (YCJJ20230576)

References

1. Ministry of Commerce of the People's Republic of China (2020) Xi Jinping's speech at the general debate of the 75th session of the United Nations General Assembly 2020. Available from: https://m.mofcom.gov.cn/article/i/jyjl/m/202009/20200903003397.shtml
2. Building Energy Efficiency Research Centre, Tsinghua University (2023) China building energy efficiency annual development research report 2023 (Urban Energy Systems Theme) [M]. China Architecture & Building Press, Beijing
3. Mei H, Chen Y, Shi X et al (2023) Barriers and strategies of building carbon neutral in cold regions [J]. Bull Natl Nat Sci Found China 37(03):364–371
4. Tian J, Xu S (2021) A morphology-based evaluation on block-scale solar potential for residential area in central China [J]. Sol Energy 221:332–347
5. Zhu S, Ma C (2022) Building performance optimization based on informed design concept [J]. World Architecture 03:84–91
6. Deng F, Tan H (2021) Research on energy consumption and photovoltaic substitution rate of different types of residence in shanghai based on nearly zero energy consumption [J]. Build Sci 37(04).9–18
7. Zhu L, Jing Y, Yin B et al (2023) Photovoltaic potential evaluation method based on parametricanalysis and its application [J]. Build Energy Eff 51(3):58–65+138
8. Huang X, Qu D, Sun C et al (2023) Construction of visual comfort prediction model FORDAYLIT OFICE space based on deep learning [J]. Archit J 10:50–54
9. Zhu S, Ma C, Ke X et al (2021) A rapid optimization approach to energy consumption-led buildingperformance for early design phase [J]. The Architect 06:69–76
10. East China Construction Group Co (2017) Architectural design sourcebook (Volume 3) [M]. China Architecture & Building Press, Beijing
11. Li G, Sun C, Sang M et al (2024) The synergistic effect of multiple design factors on building energy consumption of office blocks: a case study of Wuhan, China [J]. J Build Eng 95:110200
12. Xu S, Jiang H, Xiong F, Zhang C, Xie M, Li Z (2021) Evaluation for block-scale solar energy potential of industrial block and optimization of application strategies: a case study of Wuhan, China [J]. Sustain Cities Soc (72)
13. Liao W, Heo Y, Xu S (2019) Simplified vector-based model tailored for urban-scale prediction of solar irradiance [J]. Sol Energy 188:483
14. Xie M, Wang M, Zhong H et al (2023) The impact of urban morphology on the building energy consumption and solar energy generation potential of university dormitory blocks [J]. Sustain Cities Soc 96:104644
15. Jiang Q, Huang C, Wu Z et al (2024) Predicting building energy consumption in urban neighborhoods using machine learning algorithms [J]. Front Urban Rural Planning 2(6):1–21

Quantitative Analysis and Optimization Strategies on the Impact of Morphological Layout of Old Urban Residential Communities in Wuhan on Building Energy Consumption

Xinran Jiang and Yiquan Zou(✉)

Hubei University of Technology, Wuhan, China
2010621220@hbut.edu.cn, zouyq@mail.hbut.edu.cn

Abstract. In recent years, with the advancement of urban renewal, research on building energy efficiency at the urban block scale has increasingly become a hot topic. Based on the current situation of old urban residential communities in Wuhan City, this paper deeply investigates the impact of community morphological layout on building energy consumption through GIS technology and digital simulation methods. A total of 183 old urban residential communities in the main administrative districts of Wuhan City were selected as samples, and a 3D urban model was constructed to perform energy consumption simulations using the Honeybee plugin. Twelve morphological indicators, including building floor area ratio, building density, average height of individual buildings, and weighted average height of building area, were analyzed to explore their correlations with building heating energy consumption, cooling energy consumption, and total energy consumption. Through Pearson correlation analysis and ridge regression prediction models, it was found that specific morphological indicators significantly impact building heating, cooling, and total energy consumption. Based on the research results, this paper proposes strategic suggestions for optimizing the morphological layout of old urban residential communities in Wuhan City, aiming to improve energy efficiency, reduce building energy consumption, and thereby promote sustainable urban development. This study not only provides scientific data support for urban designers but also offers an important reference for future renovation of old urban residential communities.

Keywords: Wuhan City · Old urban residential communities · Morphological layout · Building energy consumption · Digital simulation · Pearson correlation analysis · Ridge regression prediction model · Optimization strategies

Funded Projects: Major Program (JD) of Hubei Province (2023BAA007).

J. Fu et al. (eds.), *AI for Architecture*, Computational Design and Robotic Fabrication,
https://doi.org/10.1007/978-981-95-0974-4_12

1 Introduction

1.1 Research Background

China's rapid development has brought about urban issues. Large-scale urbanization has led to a significant increase in energy consumption during the construction, operation, and demolition phases of buildings, which in turn causes carbon emissions and environmental pollution. Rising outdoor temperatures contribute to the heat island effect, impacting residents' physical and mental health. Currently, China has entered the mid-to-late stage of urbanization, and stock renewal has gradually become a vital pathway for urban transformation and development. Among them, old urban residential communities not only occupy the highest proportion of land use in urban built-up areas but are also closely related to people's daily lives. Therefore, the renewal of old districts is inevitably an important direction for urban renewal, and research perspectives have shifted from individual building energy consumption to urban-scale energy consumption.

Foreign scholars have conducted extensive research on building energy consumption at the block scale. Taylor [1] proposed a method to analyze urban building energy consumption by leveraging urban data from geographic information systems (GIS) and combining it with the urban spatial morphology of Leicester, UK. They also developed a model to analyze the geographical distribution of energy consumption in urban spaces, revealing the energy consumption characteristics of local non-residential buildings. Quan's et al. [2] study in downtown Portland, USA, found that enclosed blocks exhibited optimal energy performance when building density was below 50%; as density increased, point-style blocks performed better in terms of energy consumption.

Many domestic scholars have also conducted corresponding research on the energy consumption of residential blocks in different climatic regions. For instance, Yuan Qing, Zhao Yan, Leng Hong, and their team [3] studied small-town residential blocks in Changxing County, Zhejiang Province. They divided 17 architectural prototypes based on five typical floor area ratios and conducted energy consumption simulations, deriving corresponding planning strategies and block patterns. Yi and Jinxuan [4] studied the energy performance of high-FAR morphological types in Shanghai, extracted morphological design parameters (block orientation, group combination, and window-to-wall ratio) of residential areas, and quantitatively analyzed their relationships with building energy consumption, summarizing the morphological characteristics of energy-efficient high-FAR residential areas.

Furthermore, with the development of digital technology, an increasing number of researchers are employing digital simulation technology, which offers higher cost-effectiveness, flexibility, controllability, and precision compared to actual measurements, facilitating control and optimization. Oh and Kim [5] investigated the impact of block morphological indicators on building energy performance. Based on geometric indicators from three districts in South Korea, they identified block types and characteristics, selected five urban indicators, and simulated the unit area energy consumption values for cooling, heating, and lighting of 100 buildings in the districts. Through regression analysis, they identified significant influencing indicators. Finally, they classified 13 block types from eight urban clusters using a genetic clustering algorithm, deriving the characteristic indicators of different block types. Sarralde et al. [6] optimized the morphological parameters of residential blocks in London to enhance the solar potential of building roofs and facades, analyzing and studying the relationship between urban morphology and solar potential to improve urban solar energy utilization.

Researchers have also conducted energy consumption research and predictions by establishing idealized block models. Baiyang [7], using Harbin's severe cold climate as a backdrop, simulated three residential forms with the same FAR but different block layouts (detached, three-sided street-facing, and enclosed) to analyze the solar radiation conditions on various building facades at different times. Vermeulen et al. [8], based on meteorological data from Paris, employed evolutionary algorithms to explore the solar radiation potential of blocks composed of buildings of varying heights, discovering that pyramid- and courtyard-style block height distribution patterns offered better radiation collection efficiency than uniform height distribution patterns.

However, due to the differences between idealized and actual blocks, research findings may deviate from reality, limiting the scope of studies. With the widespread application of GIS technology, more researchers are leveraging satellite data to build three-dimensional urban models, enabling simulations and calculations of real-world cities through digital methods. Combining real blocks with simulation calculations can circumvent the limitations of both actual surveys and idealized block studies, leading to better research conclusions.

1.2 Research Method

Taking Wuhan City, Hubei Province as an example, this paper acquired the shp and dbf files of Wuhan through GIS technology and established an urban model of Wuhan using the Grasshopper parametric platform.

To explore the impact of morphological indicators on the building energy consumption in old urban residential communities, this paper identified 183 target blocks through surveys of actual neighborhoods in Wuhan. This paper will calculate the heating energy consumption, cooling energy consumption, and total energy consumption of the 183 blocks.

According to numerous related studies, it can be found that there is a strong correlation between block morphology and block energy consumption environment. Jia [9] discovered that building density and building height are related to the potential for photovoltaic utilization in blocks; Chuan [10] found that building density has a significant correlation with the energy consumption of point-type and linear residential buildings; Zhixin [11] discovered that reducing the shape coefficient also reduces the heating and cooling energy consumption of buildings.

This study takes single-unit area, standard deviation of single-unit area, standard deviation of building height, building density, weighted average height of building area, average building height, average building distance, average building perimeter, average shortest distance between buildings, floor area ratio, average difference in building perimeter, and standard deviation of the shortest distance between buildings as the morphological indicators for research. The study aims to investigate the correlation and obtain patterns by examining the morphological indicators and building energy consumption of the 183 blocks.

1.3 Data Processing

This study aims to explore the correlation between urban block morphological indicators and building energy consumption through extensive data analysis and summarize the optimal ranges of these morphological indicators, thereby proposing corresponding design strategies.

Firstly, the study employs correlation analysis methods to investigate the influence patterns of block morphological indicators on the objective functions.

Pearson correlation analysis has important application value in the analysis of block morphology indicators and building energy consumption. By calculating the correlation coefficient between block morphology indicators and building energy consumption, the impact of these indicators on building energy consumption can be quantitatively evaluated, providing a scientific basis for us to formulate more effective optimization strategies.

However, Pearson correlation analysis also possesses some limitations. Therefore, the study will adopt a regression model approach to screen for urban morphological indicators that are relevant to the objective function, while simultaneously analyzing how these indicators influence the objective function. The study will establish a regression prediction model using the Linear Ridge Regression algorithm.

This study aims to explore the relationship between morphological indicators and energy consumption in urban blocks, and summarize urban design guidelines for urban designers. Designers can use these guidelines to carry out relevant urban designs.

2 Research Method for Block Form Based on Digitization and Artificial Intelligence

This paper takes Wuhan City as an example. Wuhan, the capital city of Hubei Province, has a permanent resident population of approximately 13,774,000. Wuhan City comprises multiple administrative districts, including Jiang'an District, Jianghan District, Qiaokou District, Hanyang District, Wuchang District, Qingshan District, Hongshan District, Dongxihu District, Caidian District, Jiangxia District, Huangpi District, Xinzhou District, as well as Wuhan Economic and Technological Development Zone, Donghu New Technology Development Zone, and Donghu Eco-tourism Scenic Area. Utilizing GIS technology, this paper has captured partial map information of Wuhan's central urban area, processed it into a planar representation of building contours, and assigned heights to individual blocks, ultimately resulting in a regional model that serves as foundational research data. The primary research objects in this paper focus on four districts: Jianghan District, Hanyang District, Wuchang District, and Hongshan District. As the main administrative districts of Wuhan, they boast a long history, concentrating a significant portion of Wuhan's population. Simultaneously, these districts harbor numerous old urban residential communities, causing numerous inconveniences for residents. Consequently, these areas are progressively implementing the renovation of old urban residential communities.

2.1 Introduction of Block Form Indicators

Some studies have conducted quantitative analysis of block spatial morphology through morphological indicators, and scholars from different backgrounds have proposed a large number of related indicators. Among these indicators, some have close theoretical connections and can be converted through formulas. The study requires comparing the indicator values across numerous blocks to screen out indicators with higher eigenvalues and establish an effective spatial morphology quantification index system. This study selected 12 indicators, which are divided into two aspects: 1. Block Morphological Indicators (2 indicators); 2. Building Morphological Indicators within the Block (10 indicators). Table 12.1 provides a detailed description of the 12 selected indicators

Table 12.1 Twelve morphological indicators selected for the study (*Source* self drawn by the author)

Indicator name	Indicator type	Indicator symbol	Formula	Indicator unit
Plot building floor area ratio	Overall indicators of block morphology	FAR	$FAR = \frac{\sum_{i=1}^{n}(A_i \times N_i)}{TA}$	/
Plot building density		BCR	$BCR = \frac{\sum_{i=1}^{n} A_i}{TA}$	/
Average height of individual buildings	Morphological indicators of buildings within the block	BH_{ave}	$BH_{ave} = \frac{\sum_{i=1}^{n} BH_i}{n}$	m
Standard deviation of average height of individual buildings		BH_{sd}	$BH_{sd} = \sqrt{\frac{\sum_{i=1}^{n}(BH_i - BH_{ave})^2}{n-1}}$	m
Weighted average height of building area		WAHBA	$WAHBA = \frac{\sum_{i=1}^{n}(H_i \times S_i)}{\sum_{i=1}^{n} S_i}$	m
Average area of individual buildings		BA_{ave}	$BA_{ave} = \frac{\sum_{i=1}^{n} BA_i}{n}$	m^2
Standard deviation of average area of individual buildings		BA_{sd}	$BA_{sd} = \sqrt{\frac{\sum_{i=1}^{n}(BA_i - BA_{ave})^2}{n-1}}$	m^2

(*continued*)

Table 12.1 (*continued*)

Indicator name	Indicator type	Indicator symbol	Formula	Indicator unit
Average perimeter of individual buildings		BL_{ave}	$BL_{ave} = \frac{\sum_{i=1}^{n} BL_i}{n}$	m
Standard deviation of average perimeter of individual buildings		BL_{sd}	$BL_{sd} = \sqrt{\frac{\sum_{i=1}^{n} (BL_i - BL_{ave})^2}{n-1}}$	m
Average distance between buildings		BD_{ave}	$BD_{ave} = \frac{\sum_{i=1}^{n} BD_i}{n}$	m
Average shortest distance between buildings		BD_{min}	$BD_{min} = \frac{\sum_{i=1}^{n} BD_i}{n}$	m
Standard deviation of average shortest distance between buildings		BD_{sd}	$BD_{sd} = \sqrt{\frac{\sum_{i=1}^{n} (BD_i - BD_{ave})^2}{n-1}}$	m

2.2 Introduction of Block Simulation Settings

This paper conducts a block-level energy consumption analysis using the Honeybee plugin, with a primary focus on the impact of block morphological indicators on energy consumption. Consequently, the energy consumption settings for the model refer to the ASHRAE standards, and the simulation settings and objective function settings are detailed in Table 12.4. Honeybee employs the OpenStudio core for calculations (OpenStudio is a building energy simulation software developed by the U.S. National Renewable Energy Laboratory, with contributions from multiple organizations. It integrates EnergyPlus for energy consumption simulations and Radiance for daylighting simulations, and is available for Windows, Mac, and Linux). This tool specifically calculates the building energy consumption of old urban residential communities (Table 12.2).

Table 12.2 Block simulation setting (*Source* self drawn by the author)

Simulation setting	Setting situation	Description
Climate regions	4 ASHRAE climate zone	Wuhan is a typical climate region with hot summers and cold winters
Year built	ASHRAE 90.1 2015	The buildings in the surveyed area were mostly constructed between 1980 and 2000
Building function	Residential area	Residential Block
Building structure type	Steel framed	The buildings in the surveyed area are mostly multi story residential buildings
Research objective function	Unit	Description
$E_{Heating}$	$kW \cdot h/m^2$	Building heating energy consumption
$E_{cooling}$	$kW \cdot h/m^2$	Building cooling energy consumption
E_{Total}	$kW \cdot h/m^2$	Total energy consumption of buildings

2.3 Data Processing Method Setting

The study obtained a large number of combinations of block morphological indicators and corresponding objective function values through sampling. Further analysis was conducted on these data. The study first uses the Pearson correlation analysis to explore how changes in block morphology indicators affect the objective function. The Pearson correlation analysis is a statistical method that measures the strength and direction of the linear relationship between two variables. Its fundamental principle involves calculating the ratio of the covariance between two variables to the product of their respective standard deviations, resulting in a value ranging from −1 to 1, known as the Pearson correlation coefficient (r). The formula for calculating r is

$$r = \frac{\mathrm{cov}(X, Y)}{\sigma_X \sigma_Y}$$

Among them, cov(X, Y) represents the covariance between variables X and Y, and σ_X and σ_Y represent the standard deviation of variables X and Y, respectively. The value range of r is between -1 and 1, where $r > 0$ indicates positive correlation, that is, when one variable increases, the other variable also tends to increase; $r < 0$ indicates a negative correlation, meaning that when one variable increases, the other variable tends to decrease; $r = 0$ indicates that there is no linear relationship between the two variables.

Due to the non-linear correlation between some block morphology indicators and the objective function, further analysis will be conducted through regression models. Ridge Regression Analysis is a biased estimation regression method specifically designed to handle collinear data issues, essentially an improvement upon the least squares estimation method. In regression analysis, when a high degree of correlation exists among the independent variables, the traditional least squares method may lead to inaccurate estimates and large variances. To overcome this shortcoming, Ridge Regression Analysis introduces a regularization term into the loss function, which reduces the variance of the coefficient estimates, thereby enhancing the stability and predictive power of the model.

Specifically, the loss function of ridge regression consists of two parts: one is the sum of squared residuals (RSS) of ordinary least squares regression, which is used to measure the model's fit to the data; The second is the regularization term, which is used to control the complexity of the model. The regularization term usually weights the sum of squares of the coefficient vectors, with the weights determined by the regularization parameter α. Therefore, the loss function of ridge regression can be expressed as

$$L(\beta) = RSS(\beta) + \propto \times \|\beta\|^2$$

Among them, β is the coefficient vector, RSS is the sum of squared residuals, α is the regularization parameter, and $\|\beta\|^2$ is the sum of squared coefficient vectors. By adjusting the value of α, a balance can be found between the degree of fitting and the complexity of the model.

Before conducting ridge regression, linear regression is generally used first. Ridge regression is used only when the variance inflation factor (VIF) of the independent variable is found to be too large, exceeding 10. The ridge regression model requires confirmation of the K value, which is usually automatically identified using the variance inflation factor method. The K value of the ridge regression prediction model used in the study will be determined by the variance inflation factor method. The variance inflation factor can measure the severity of multicollinearity. When the VIF value is greater than 10, the model may exhibit severe multicollinearity. In this case, ridge regression can be attempted and the appropriate K value can be determined by observing the variation of VIF value with K value. Generally speaking, as the K value increases, the VIF value gradually decreases. When the VIF value drops to an acceptable range, it can be considered that a suitable K value has been found.

The general principles for selecting the K value are as follows: the ridge estimates of each regression coefficient are basically stable; the signs of regression coefficients with unreasonable estimates by the least squares method become reasonable after ridge estimation; the regression coefficients do not have absolute values that defy economic sense; and the sum of squared residuals does not increase significantly.

In the research and analysis of aging residential communities, the Ridge Regression model holds significant importance. When investigating the energy consumption of buildings in such communities, multiple influencing factors often need to be taken into consideration, among which high correlations, known as multicollinearity, may exist. Traditional linear regression models, when confronted with the issue of multicollinearity, may lead to instability in parameter estimation or even draw erroneous conclusions. The Ridge Regression model, by incorporating a penalty term into the objective function, effectively addresses this issue, rendering parameter estimation more stable and reliable. Furthermore, by adjusting the regularization parameter, the Ridge Regression model can to a certain extent mitigate overfitting, thereby enhancing the prediction accuracy of the model. In the context of studying building energy consumption in aging residential communities, this signifies that the Ridge Regression model can more precisely forecast building energy consumption under different conditions, providing a more scientific basis for energy-saving renovations and energy management. Compared to other regression models, the Ridge Regression model demonstrates notable advantages in handling collinear data, preventing overfitting, and improving model stability and generalization ability, making it a highly effective statistical method.

3 Data Analysis

3.1 Overview of Simulation Results

After simulating 183 blocks, 168 effective objective functions and morphological indicators were obtained. The study analyzed 168 sets of morphological indicators and their corresponding objective function values, and found that: 1 The value of the objective function varies due to changes in morphological indicators, with the maximum value of building heating energy consumption being 86.806 $\text{kW} \cdot \text{h/m}^2$ and the minimum value being 32.252 $\text{kW} \cdot \text{h/m}^2$, with a variation range of 62.85%; On the other hand, the maximum energy consumption for building refrigeration is 107.675 $\text{kW} \cdot \text{h/m}^2$, and the minimum is 67.106 $\text{kW} \cdot \text{h/m}^2$, with a variation of 37.68%; Finally, the maximum total energy consumption is 118.368 $\text{kW} \cdot \text{h/m}^2$ and the minimum is 65.533 $\text{kW} \cdot \text{h/m}^2$, with a variation of 44.64% (Fig. 12.1).

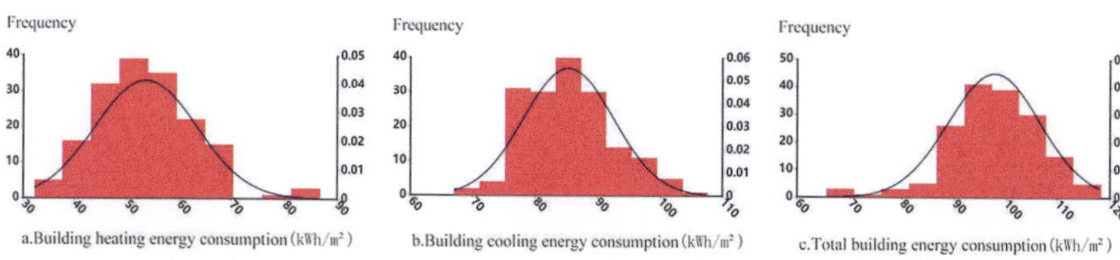

Fig. 12.1 Data overview (*Source* self drawn by the author)

3.2 Correlation Analysis Between Block Morphology Indicators and Objective Function

The study analyzed the correlation between the morphological indicators of 2016 blocks and their corresponding objective function values through Pearson correlation analysis. Research has found a strong correlation between some morphological indicators of certain neighborhoods and the objective function. Among them, there is a strong negative correlation between building heating energy consumption, building plot ratio, individual area, average perimeter of buildings, weighted average height of building area, and average height of buildings; There is a strong negative correlation between building cooling energy consumption and building plot ratio, unit area, and average building perimeter; There is a strong positive correlation between total energy consumption and the average distance between buildings. There is no strong correlation coefficient between the morphological indicators of other blocks and the objective function, indicating that the relationship between these block morphological indicators and the objective function is not linearly correlated and requires further analysis (Figs. 12.2, 12.3 and 12.4).

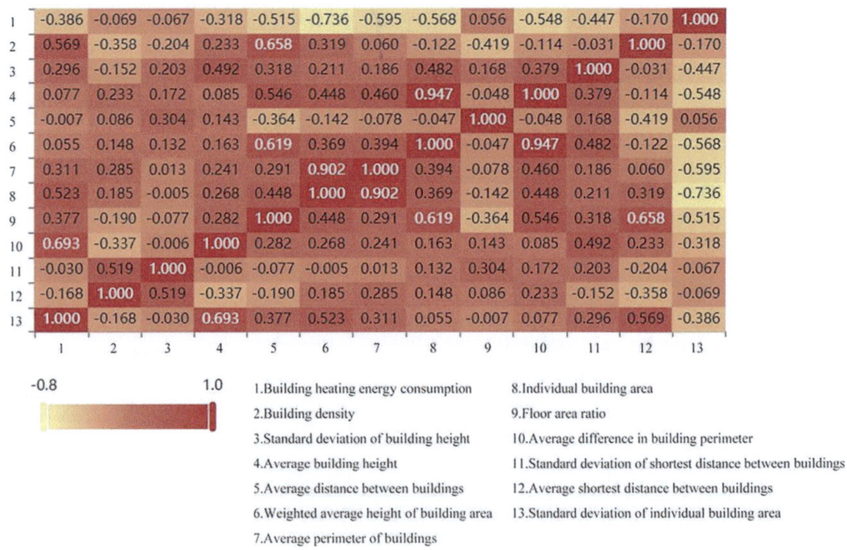

Fig. 12.2 Correlation analysis of building heating energy consumption (*Source* self drawn by the author)

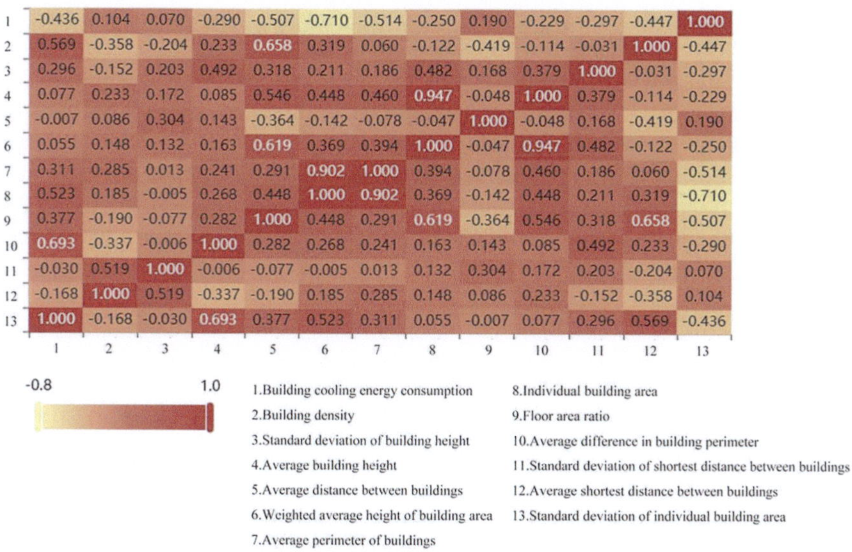

Fig. 12.3 Correlation analysis of building cooling energy consumption (*Source* self drawn by the author)

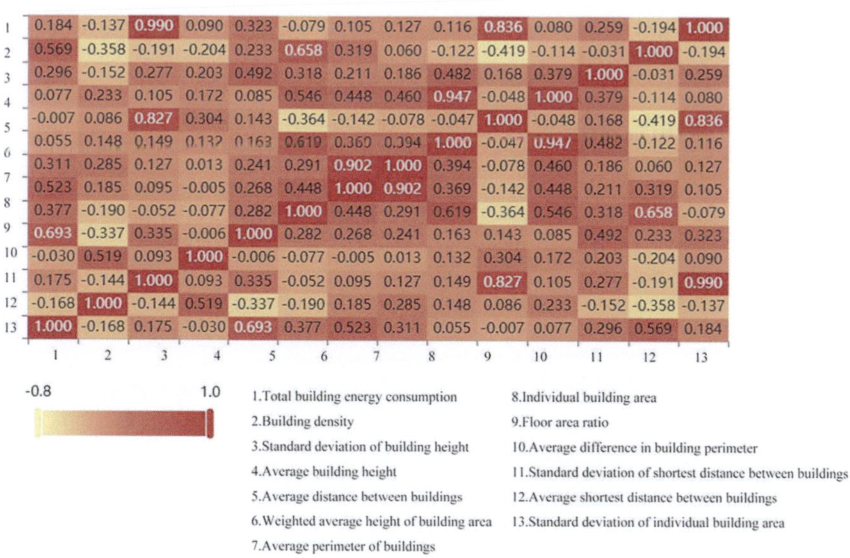

Fig. 12.4 Correlation analysis of total building energy consumption (*Source* self drawn by the author)

3.3 Prediction Model of Block Morphology Indicators for Objective Function

A ridge regression prediction model was established between the morphological indicators of the 2016 block group and various objective functions. The study found that in terms of building heating energy consumption, the R^2 value of the ridge regression model reached 0.7, in terms of building cooling energy consumption, the R^2 value of the ridge regression model reached 0.6, and in terms of total energy consumption, the R^2 value of the ridge regression model reached 0.97, indicating good model performance (Tables 12.3, 12.4 and 12.5).

Table 12.3 Ridge regression analysis results of building heating energy consumption (*Source* self drawn by the author)

K = 0.17	Non standardized coefficient		Standardized coefficient	t	P	R^2	Adjustment R^2	F
	B	Standard error	Beta					
Constant	79.756	1.623	–	49.145	0.000***	0.666	0.658	81.258(0.000***)
Building plot ratio	−1.552	0.735	−0.092	−2.111	0.036**			
Standard deviation of building height	−0.753	0.169	−0.183	−4.441	0.000***			
Unit area	−0.017	0.001	−0.496	−12.047	0.000***			
Weighted average Height of building Area	−0.586	0.127	−0.205	−4.611	0.000***			

Dependent variable: Building heating energy consumption
Note * * *, * *, * represent significance levels of 1%, 5%, and 10%, respectively

Table 12.4 Ridge regression analysis results of building cooling energy consumption (*Source* self drawn by the author)

K = 0.167	Non standardized coefficient		Standardized coefficient	t	P	R^2	Adjustment R^2	F
	B	Standard error	Beta					
Constant	101.029	1.519	-	66.524	0.000***	0.599	0.587	48.449(0.000***)
Building plot ratio	−2.173	0.585	−0.174	−3.714	0.000***			
Standard deviation of building height	−0.476	0.137	−0.156	−3.482	0.001***			

(*continued*)

Table 12.4 (*continued*)

K = 0.167	Non standardized coefficient		Standardized coefficient	t	P	R^2	Adjustment R^2	F
	B	Standard error	Beta					
Average height of buildings	0.272	0.1	0.125	2.723	0.007***			
Unit area	−0.014	0.001	−0.525	−11.322	0.000***			
Building density	−8.639	2.889	−0.133	−2.99	0.003***			

Dependent variable: Building cooling energy consumption
Note * * *, * *, * represent significance levels of 1%, 5%, and 10%, respectively

Table 12.5 Ridge regression analysis results of total building energy consumption (*Source* self drawn by the author)

K = 0.064	Non standardized coefficient		Standardized coefficient	t	P	R^2	Adjustment R^2	F
	B	Standard Error	Beta					
Constant	−579,189.261	192,498.95	–	−3.009	0.003***	0.977	0.976	1735.234(0.000***)
Area	76.558	1.724	0.744	44.418	0.000***			
Average shortest distance of buildings	−64,296.565	13,101.381	−0.058	−4.908	0.000***			
Average distance of buildings	22,705.242	1715.532	0.222	13.235	0.000***			
Unit area	1055.082	174.691	0.072	6.04	0.000***			

Dependent variable: Total building energy consumption
Note * * *, * *, * represent significance levels of 1%, 5%, and 10%, respectively

3.4 Analysis of Typical Block Morphology

To further analyze the relationship between urban block morphological indicators and objective functions, the study extracted the top 5 optimal solutions for each objective function from 168 cases, forming an optimal solution set. Simultaneously, the study also extracted the bottom 5 worst solutions for each objective function, composing a worst solution set. By individually analyzing these 30 solutions, the study found that the optimal solutions exhibited certain similarities, and similarly, the worst solutions also shared some common characteristics. This indicates that designers can summarize the block architectural forms of both optimal and worst solutions to derive relevant design guidelines.

Regarding heating energy consumption, a comparison between the optimal and worst solution sets revealed that larger individual building area, standard deviation of building height, floor area ratio (FAR), and weighted average height of building area all contributed to lower heating energy consumption. This suggests that, to a certain extent, as the individual building area increases, the relative area of its envelope structure may decrease, meaning that less heating energy is required per unit volume of the building. This is because larger buildings may exhibit better economies of scale in terms of insulation, reducing heat loss. Additionally, higher FAR and standard deviation of building height lead to a more compact layout, which facilitates the creation of a better microclimate, such as reducing wind speed and increasing solar radiation, both beneficial for heat retention (Figs. 12.5, 12.6 and Table 12.6).

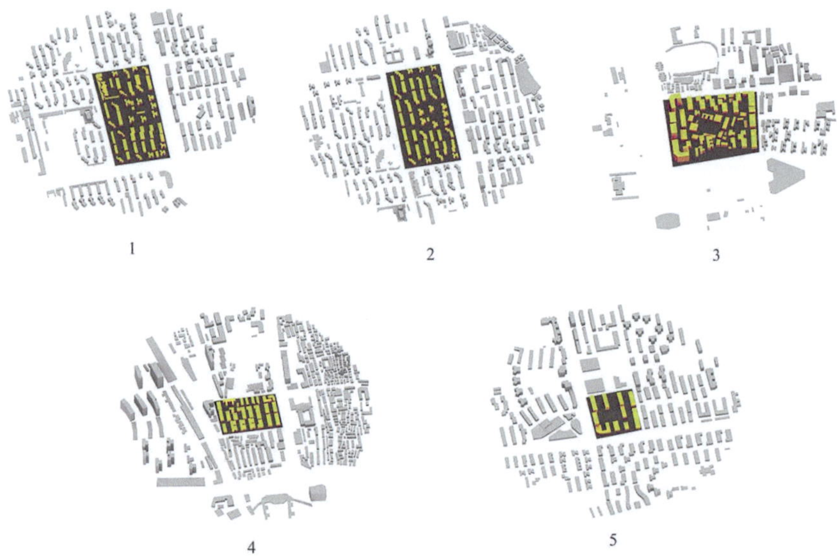

Fig. 12.5 The optimal solution set for building heating energy consumption (*Source* self drawn by the author)

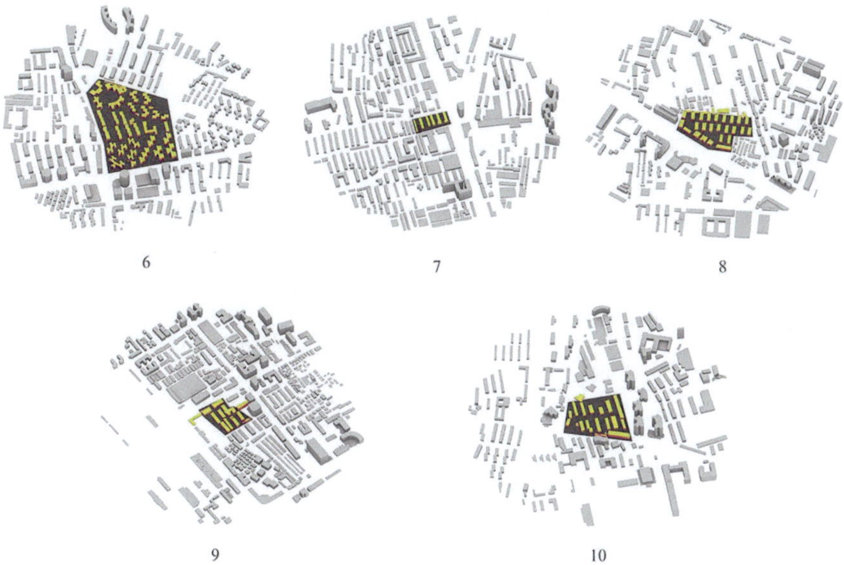

Fig. 12.6 The worst solution set for building heating energy consumption (*Source* self drawn by the author)

Table 12.6 Typical cases of building heating energy consumption (*Source* self drawn by the author)

	Number	Building heating energy consumption ($kW \cdot h/m^2$)
The optimal solution	1	32.2520
	2	34.9410
	3	37.0540
	4	37.5720
	5	37.5840
The worst solution	6	70.2750
	7	79.0500
	8	83.0840
	9	85.5400
	10	86.8050

In terms of cooling energy consumption, a comparison between the optimal and worst solution sets shows that higher floor area ratio (FAR), standard deviation of building height, individual building area, and building density all lead to lower cooling energy consumption. Conversely, lower average building height is also associated with reduced cooling energy consumption. This indicates that a more compact layout and larger individual building area reduce the heat radiation received by buildings and the ground. Lower-rise buildings, on the other hand, are more adept at reducing indoor temperatures through natural ventilation and radiative cooling in summer. This is because their roofs and external wall areas are relatively smaller, receiving less solar radiation. Additionally, low-rise buildings are more susceptible to the cooling effect of ground radiation, which helps to lower indoor temperatures and thus reduce cooling energy consumption (Figs. 12.7, 12.8 and Table 12.7).

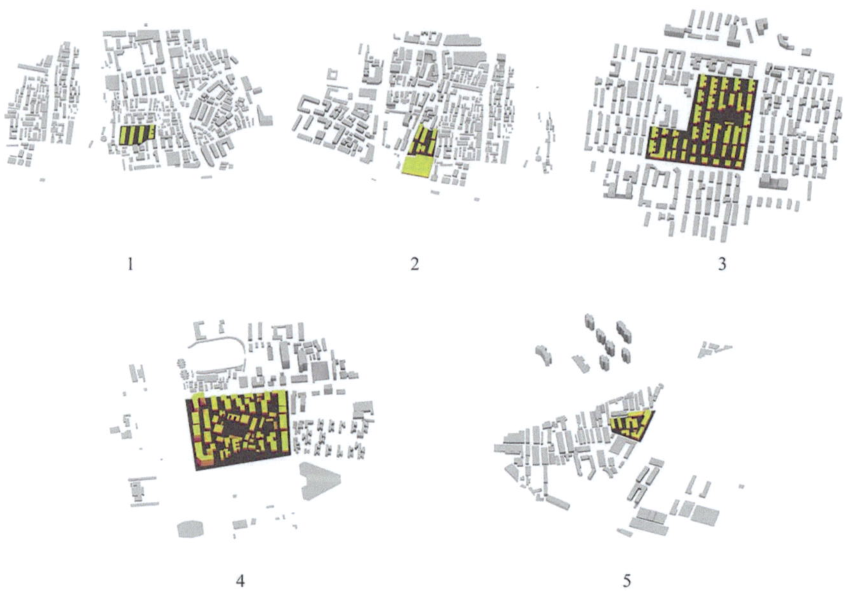

Fig. 12.7 The optimal solution set for building cooling energy consumption (*Source* self drawn by the author)

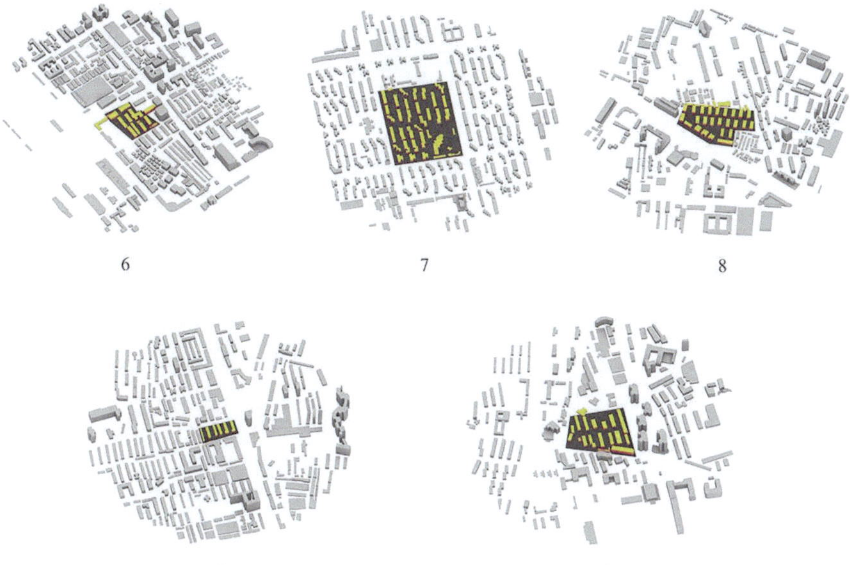

Fig. 12.8 The worst solution set for building cooling energy consumption (*Source* self drawn by the author)

Table 12.7 Typical cases of building cooling energy consumption (*Source* self drawn by the author)

	Number	Building cooling energy consumption (kW · h/m^2)
The optimal solution	1	67.1060
	2	70.9140
	3	73.6120
	4	73.8010
	5	74.6140
The worst solution	6	102.2290
	7	102.3280
	8	103.3480
	9	103.5830
	10	107.6750

Regarding total energy consumption, a comparison between the optimal and worst solution sets reveals that smaller average distances between buildings and a smaller total area of the region lead to lower total energy consumption. This suggests that a more compact layout contributes to creating a favorable microclimate, thereby reducing the overall energy consumption of buildings (Figs. 12.9, 12.10 and Table 12.8).

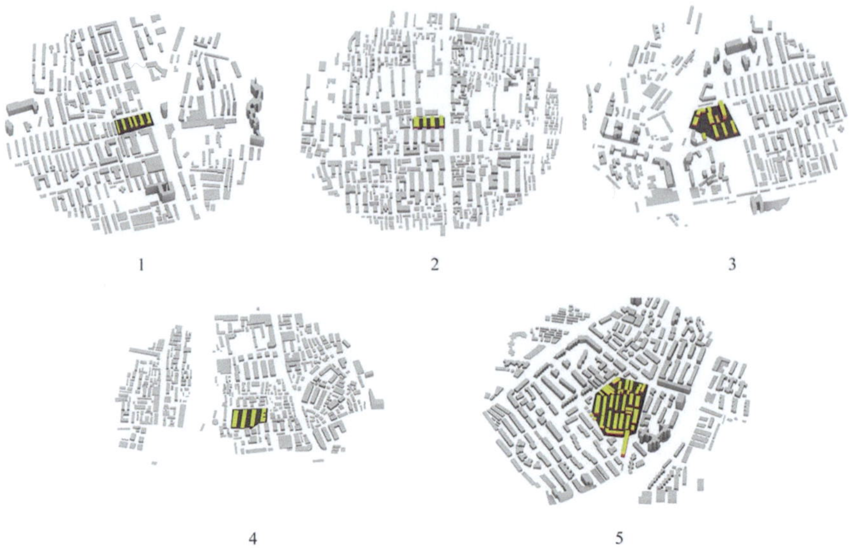

Fig. 12.9 The optimal solution set for total building energy consumption (*Source* self drawn by the author)

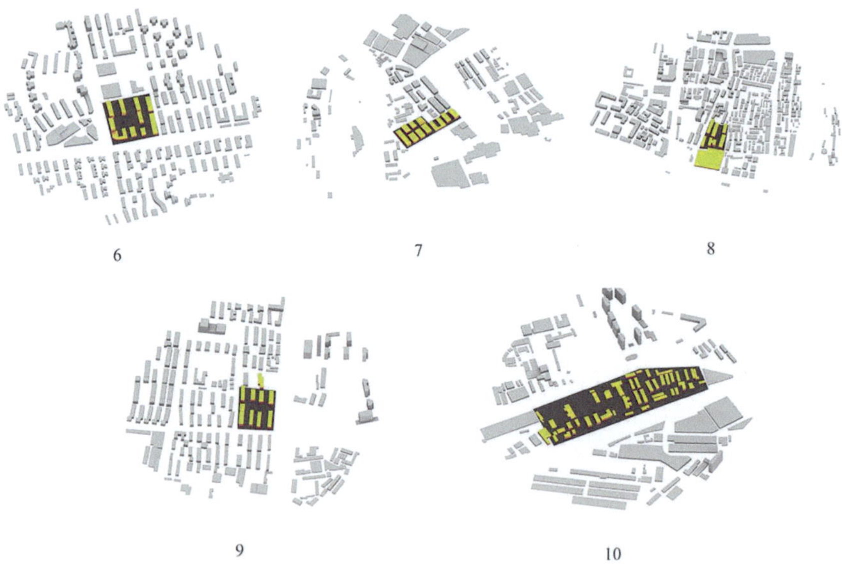

Fig. 12.10 The worst solution set for total building energy consumption (*Source* self drawn by the author)

Table 12.8 Typical cases of total building energy consumption (*Source* self drawn by the author)

	Number	Total building energy consumption ($kW \cdot h/m^2$)
The optimal solution	1	65.5331
	2	70.3067
	3	70.7937
	4	74.3444
	5	77.5056
The worst solution	6	114.4262
	7	114.8231
	8	115.3210
	9	117.2420
	10	118.3683

4 Conclusion and Prospect

This study, through an in-depth analysis of energy consumption in old urban residential communities in Wuhan's main urban area, reveals the significant impact of key parameters such as individual building area, standard deviation of building height, floor area ratio, weighted average height of building area, building density, and average building height on energy consumption. Specifically, in terms of heating energy consumption, large-scale buildings exhibit lower energy consumption characteristics due to their scale effect in thermal insulation and the favorable microclimate environment created by their compact layout. For cooling energy consumption, a compact layout combined with larger individual building areas effectively reduces heat radiation reception, and when integrated with the natural ventilation and radiant heat dissipation advantages of low-rise buildings, significantly lowers the cooling demand of buildings. Comprehensive analysis of total energy consumption indicates that a more compact neighborhood layout not only enhances microclimate optimization but also notably reduces overall building energy consumption.

Regarding the design strategies and guidelines for the renovation of old residential communities and the development of new residential blocks, this study proposes the following substantive suggestions:

1. Optimize individual building design: Encourage the appropriate increase in individual building area during renovation to reduce the relative area of envelope structures and improve thermal insulation performance. Simultaneously, focus on enhancing the quality of insulation materials for critical components such as exterior walls and roofs to minimize heat loss.
2. Compact layout and rational planning: Prioritize compact and orderly layouts in community and design renovation planning. Rationally adjust FAR, standard deviation of building height, and building density to promote microclimate improvement.

Additionally, avoid excessive density that could hinder ventilation, ensuring smooth airflow within the neighborhood.

3. Rational distribution of high- and low-rise buildings: Where feasible, configure high- and low-rise buildings in residential blocks strategically to leverage the shading effects between buildings, fostering a favorable microclimate.

As urban renewal progresses and the focus shifts to stock renovations, future research could further explore energy consumption optimization strategies for old community renovations under different climatic conditions, as well as assess the economic and social benefits of different renovation schemes. Moreover, with technological advancements and the application of new materials, it is crucial to continuously monitor and introduce more advanced energy-saving technologies and products to bolster the renovation of old communities and reduce block-level energy consumption. Furthermore, strengthening interdisciplinary collaboration and exchange will jointly drive sustainable urban development and the achievement of energy conservation and emission reduction targets.

References

1. Taylor S, Fan D, Rylatt M (2014) Enabling urban-scale energy modelling: a new spatial approach [J]. Build Res Inf 42(1):4–16
2. Quan SJ, Economou A, Grasl T et al (2014) Computing energy performance of building density, shape and typology in urban context [J]. Energy Procedia 61:1602–1605
3. Hong L, Xi C, Yanhong M (2020) Research progress and enlightenment on the impact of urban form on building energy consumption [J]. J Archit (2):7
4. Yi W, Jinxuan W (2020) Research on energy consumption performance of high floor area ratio residential communities in hot summer and cold winter regions: a case study of Shanghai City [J]. Archit Techniques (7):8
5. Oh M, Kim Y (2019) Identifying urb an geometric types as energy performance patterns [J]. Energy Sustain Dev, 48
6. Sarralde JJ, Quinn DJ, Wiesmann D, Steemers K (2015) Solar energy and urb an morp hology: Scenarios for increasing the renewable energy potential of neighbourhoods in London [J]. Renew Energy, 73
7. Baiyang (2017) Research on the optimization of residential form in severe cold regions based on the potential of solar energy utilization [D]. Harbin Institute of Technology
8. Vermeulen T, Merino L, Knopf-Lenoir C et al (2018) Periodic urban models for optimization of passive solar irradiation [J]. Sol Energy 162:67–77
9. Jia T (2020) Research on the potential of solar photovoltaic utilization in urban residential blocks: a case study of Wuhan City [D]. Huazhong Unive Sci Technol
10. Chuan S (2019) Study on the impact of block scale urban form on energy consumption of building clusters based on software simulation: a case study of residential buildings in Nanjing [D]. Southeast Univ
11. Zhixin L (2021) Study on the impact of urban block form layout on building energy consumption in hot summer and cold winter regions: a case study of Wuhan Blocks [D]. Huazhong Univ Sci Technol

Computational Urban Design

Prediction of Urban Traffic Accidents and Designer-Friendly Optimization Strategies

Xinning He[1], Yinan Wu[1], and Hao Zheng[2(✉)]

[1] School of Architecture, Tsinghua University, Beijing, China
{he-xn20,wuyn20}@mails.tsinghua.edu.cn

[2] Department of Architecture and Civil Engineering, University of Hong Kong, Hong Kong, China
hazheng@cityu.edu.hk

Abstract. Predicting traffic accidents plays an important role in improving transportation efficiency and urban safety. Among other things, the built environment and road network design of a city can affect the incidence of traffic accidents. Existing studies tend to focus on abstract theories, posing challenges for urban designers to apply intuitively. This study proposes a workflow to analyze and predict urban traffic accidents by integrating urban road networks, land use, and building profiles. We extracted data on traffic accident occurrences in San Francisco in 2016–2023 and mapped the coordinates to the city's land use and road network maps, and developed a graph-born graph prediction model for urban traffic accidents using GAN neural networks. The model was able to produce fairly accurate predictions of traffic accidents in the city of San Francisco. We used the model to analyze the corresponding road safety situations under common urban prototypes and road network patterns from the perspective of urban design (road modeling), and summarized the impacts of common road and land use patterns on traffic safety in the city of San Francisco, as well as the possible ways to improve them. In addition, the model is applied to other cities in the U.S. to validate the model's migration capability.

Keywords: Traffic safety · GAN · Urban design · Traffic accidents prediction

1 Introduction

Traffic safety is a critical component of public health and urban planning. The World Health Organization (WHO) reports that around 1.19 million people die annually due to road traffic crashes. Understanding the factors that contribute to traffic accidents is essential for improving urban transportation safety and efficiency. Traditionally, the focus has been on road networks and traffic flow, but the broader built environment—encompassing building layout, street width, intersection design, and land use—also plays a significant role. The road network not only facilitates city transportation but also shapes the interactions between various sites, making it crucial to view it through

Xinning He and Yinan Wu contributed equally to this manuscript.

J. Fu et al. (eds.), *AI for Architecture*, Computational Design and Robotic Fabrication,
https://doi.org/10.1007/978-981-95-0974-4_13

the lens of urban design [3]. Existing research often approaches the issue abstractly, either summarizing causes of accidents [11] or developing predictive models based on various factors [26]. While these methods offer insights, they frequently lack direct applicability for urban designers, who require intuitive guidance on how specific design choices influence traffic safety.

Studies have shown that built environment variables significantly impact traffic accidents, with land use mix (14.29%), freeway and arterial road density (12.43%), and secondary and local road density (11.54%) being the top contributors [28]. However, the lack of dynamic visualization tools to represent the relationships among these factors remains a challenge. Addressing this gap is essential for providing urban designers with clearer, more intuitive guidance that integrates architectural knowledge with traffic studies, ultimately leading to informed decisions that promote safer urban spaces.

2 Related Work

2.1 Built Environment and Traffic Accidents

Traffic safety is vital for sustainable urban development, yet the causes of traffic accidents are often complex and multifaceted. Haddon identified three decisive factors influencing road traffic safety: human factors, environment, and vehicle [9]. Insufficient oversight and enforcement of traffic regulations frequently contribute to road accidents [7]. In addition to human factors and immutable conditions like weather [2] and time [22], the built environment—including human-made structures, public infrastructure, and land use—substantially affects traffic accident occurrences of a city also has a substantial impact on the occurrence of traffic accidents [17].

The layout and design of the built environment directly influence traffic flow and safety [13, 21]. Road design, including considerations for all road users, significantly impacts road safety [1]. Research indicates that road density affects safety, with traditional views suggesting that wider, lower-density networks enhance safety [15, 27]. Conversely, evidence suggests that in dense urban areas, "less forgiving" designs—such as narrow lanes and traffic-calming measures—may improve safety [6]. Land use also significantly impacts traffic accidents, with mixed-use developments correlating with higher pedestrian collision risks [14, 24]. Multi-factorial cross-analysis is crucial for a comprehensive understanding of traffic accidents.

2.2 Prediction Models for Traffic Accidents

Traffic accidents have been a longstanding focus of road safety research. With advances in data acquisition and analysis, research has shifted from traditional statistics to large-scale, data-driven predictive models. Accident data encompasses various factors, such as urban road geography, traffic volume, weather conditions, and driver behavior, forming the basis for analysis. Data sources include government agencies, public databases, embedded road infrastructure devices, in-vehicle sensors, and social media platforms [4, 8].

Accurately identifying and predicting future accidents holds significant practical value. Current mainstream analytical methods include Bayesian networks, genetic algorithms, artificial neural networks, deep learning algorithms and others [4, 5, 8]. Traditional probabilistic methods like Decision Tree algorithms are notable for their ability to predict traffic accident severity without requiring prior probabilistic knowledge of the studied phenomena [29–31]. However, their reliance on binary classification makes it difficult to incorporate visualized information as input or output. Increasing research attention has shifted toward deep neural networks, including LSTM, CNN and GAN models, which have been successfully applied to analyze traffic speed and accident data across both spatial and temporal dimensions [32–34]. Recent advancements in large language models (LLMs) have facilitated the development of agent-based models for accident analysis and prediction, incorporating effectors, sensors, communication, intention, motivation, and cognition. These methods are primarily applied in complex systems [35], like autonomous driving [36, 37] and traffic control [38]. As this study focuses on built environment factors and excludes dynamic variables, their use is less relevant here.

2.3 Applications of GAN Models in Urban Studies

Generative Adversarial Networks (GANs), as advanced deep learning models, have shown significant potential in urban studies. GANs consist of a generator and a discriminator, which undergo adversarial training to produce data resembling real datasets. Various studies have employed GAN models for urban and transportation-related predictions. These models, such as TrafficGAN for traffic flow prediction [39], Curb-gan for spatiotemporal data analysis [34], and other GAN-based approaches for urban planning changes [40] and traffic accident prediction [41–43], have shown great value in their respective applications. Pix2Pix [44], a model suited for image translation tasks requiring paired data, has been widely applied in scenarios such as architectural plan generation [45, 46] and crime prediction [47]. More importantly, this method, which leverages real-world urban morphology image data for training and visualization, enables designers and the public to easily comprehend and apply the relationships between various factors and data.

However, existing research lacks a focus on understanding the relationship between the built environment and traffic accidents, instead prioritizing model architecture optimization. There is a notable gap in the application-level analysis, particularly in interpreting the "black box" of artificial neural networks (ANNs) from urban and architectural perspectives. Previous studies, such as the Urban-GAN model proposed by Quan et al. [48], have created multi-dimensional urban design computation systems that offer valuable insights for public participation in urban design. Our research aims to integrate expertise in architecture and urban planning, focusing on how urban-scale elements contribute to traffic safety. By using land use types and road network characteristics as key input variables, we seek to understand their influence on accident occurrences, ultimately providing designer-friendly guidance for urban traffic planning, fostering informed decision-making.

3 Methodology

3.1 Data Processing

We selected San Francisco, California, as the data source due to its rich urban environmental data, high level of urban development, and frequent traffic safety incidents. We used the open-source geographic information system QGIS to visualize the data.

The research concluded that a consecutive period of 5–10 years is the preferred time period for road accident prediction [18]. We collected two datasets of traffic accident records spanning from 2016 to 2023 from Kaggle [20] and the San Francisco government website, resulting in a total of 39,000 de-duplicated accident records. The dataset includes precise coordinates (latitude and longitude), date and time of the incidents, severity of the accidents, and the types of vehicles involved (automobile, bicycle, or pedestrian). To clearly identify the locations of the accidents, we visualized each accident as a black dot with 20% transparency and a diameter of 5 m. The image was then color-inverted to produce a black background with white dots, which served as the output image for training the GAN model.

We integrated road network data from OpenStreetMap and 2020 land use data from DataSF to be used as input for the GAN model. To improve the model's accuracy, we adjusted the RGB values, setting the R, G, and B dimensions to 0, 125, and 255, respectively. This adjustment maximizes the contrast between different color layers, thereby enhancing the model's performance and robustness (specific values are illustrated in the figure) (Fig. 13.1).

	Land Type	R	G	B
1	CIE = Cultural.institutional. Educational	125	255	0
2	MED = Medical	125	0	125
3	MIPS = Office (Management, Information, Professional Services)	125	125	0
4	MIXED = Mixed Uses (Without Residential)	0	125	0
5	MIXRES = Mixed Uses (With Residential)	125	0	255
6	PDR = Industrial (Production, Distribution, Repair)	255	125	0
7	RETAIL/ENT = Retail.Entertainment	255	0	0
8	RESIDENT = Residential	0	0	255
9	VISITOR = Hotels/Motels	125	125	255
10	VACANT = Vacant	0	125	125
11	ROW = Right-of-Way	0	255	255
12	OPENSPACE = Open Space	0	255	0
13	Missing data	255	0	255
14	Others	255	255	125
15	building	255	255	255
16	road	255	255	0

Fig. 13.1 The RGB values of different land use

Dataset Preparation

In determining the dimensions of the training images, we considered that traffic accidents often extend beyond a single intersection and may be influenced by the surrounding built environment across several blocks. Therefore, we selected an area coverage of approximately 600 m × 600 m per image. To ensure image clarity and to retain the maximum amount of useful information, we specified each image to be 512 × 512 pixels. Consequently, we selected a usual architect's scale at 1:50, and each final image represents 677.5 m × 677.5 m (see Figs. 13.2, 13.3). In order to ensure the accuracy of the model's understanding of road widths and densities, it is essential to keep the scale of the input images consistent.

Fig. 13.2 The processed map of San Francisco

We used Python Image Library (PIL) for the automated cropping of input and output images. To mitigate the potential impact of image boundaries on data analysis, we employed an overlapping strategy for adjacent images to ensure full coverage of the map area. Specifically, each subsequent image was shifted by 256 pixels either in the X or Y direction from the previous image, covering the entire area. Additionally, a filtering mechanism was implemented to exclude blank images lacking data information, thereby enhancing the efficiency of model training.

After processing, a total of 1,212 image sets were obtained. We randomly selected 1,100 image sets for the training dataset and reserved 112 image sets for the test dataset to evaluate the training results (Fig. 13.4).

Fig. 13.3 The map of traffic accidents recorded between 2016 and 2023 in San Francisco

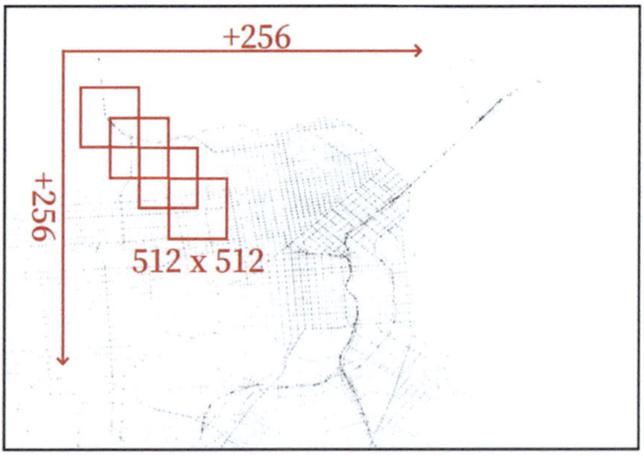

Fig. 13.4 The generation of dataset from a single large-scale map

3.2 Training

In this paper, the model training utilized the NVIDIA open-source pix2pixHD code [23], which incorporates a novel adversarial loss function along with a multi-scale generator and discriminator architecture. This setup enables the generation of high-resolution, realistic images with dimensions of 2048 × 1024 pixels. We believe this model effectively meets the image learning requirements of this study. Various hyperparameters, including learning rates, were fine-tuned multiple times to enhance training optimization. The neural network architecture of the model is illustrated in Fig. 13.5.

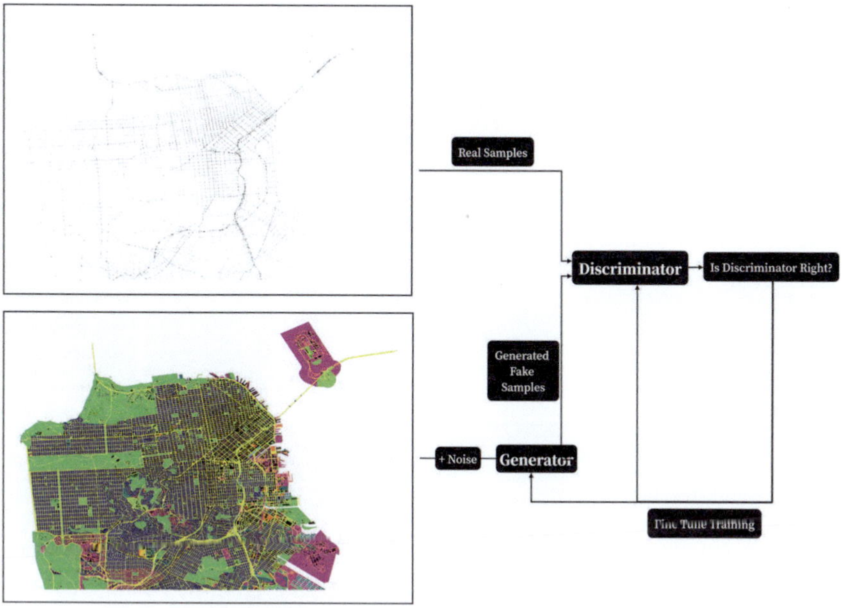

Fig. 13.5 The workflow of our model training

To achieve an optimal model, we conducted multiple rounds of hyperparameter tuning, focusing primarily on comparing learning rate, number of epochs and batch size. We ultimately determined the locally optimal learning rate to be 0.00002, conducted a total of 300 epochs of training, setting the niter value to 210.

4 Result Analysis

4.1 Training Results

The results showed a significant reduction in both D_Loss and G_Loss after training, indicating that the model achieved satisfactory performance (Fig. 13.6). As shown in Fig. 13.7, with the increase in training time and epochs for the generative adversarial network (GAN) model, the simulated traffic accident data points generated by the model progressively approximated the real images. After completing the final training, we evaluated the reserved 112 test images (Fig. 13.8).

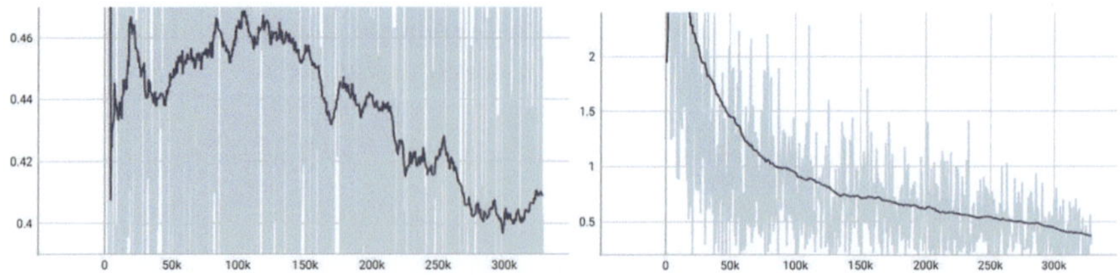

Fig. 13.6 The D_Loss and G_Loss show a decline during the training process

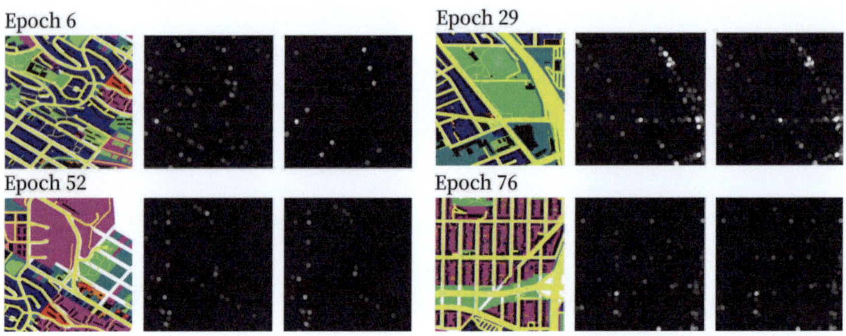

Fig. 13.7 The prediction ability strengthens with epochs

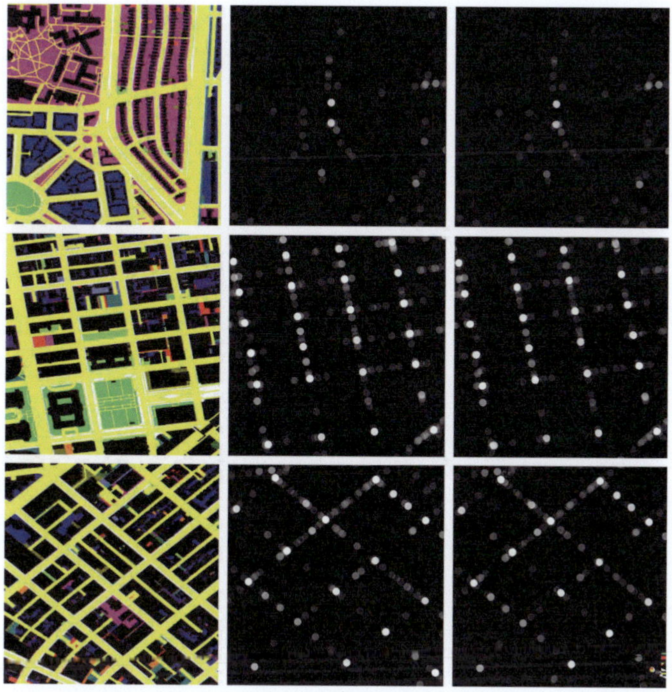

Fig. 13.8 Our model shows good quality on traffic accidents prediction. Left: input map; Middle: real accidents map; Right: generated accidents map

To assess the practical predictive accuracy of the trained model, we utilize the Fréchet Inception Distance (FID) score, which measures the diversity and quality of the generated images [10]. A lower FID score indicates that the generated images are more similar to the real ones, reflecting better model validity. Our FID value is 20.01, illustrating a positive simulation.

4.2 Traffic Accident Patterns

We began by analyzing and understanding the impact of the built environment on the number of traffic accidents using existing real-world data. We classify the patterns of road and land use relationships into the following categories: Roads (Grid Network, Curved Roads, Multiple Road Intersections, Sparsely Built Area) and Land Use Types (Residential Dominant, Commercial/Entertainment Dominant, Mixed Land Uses) (see Fig. 13.9). In this study, sparsely built area are not observed in residential-dominant areas, but in greenplace-dominant.

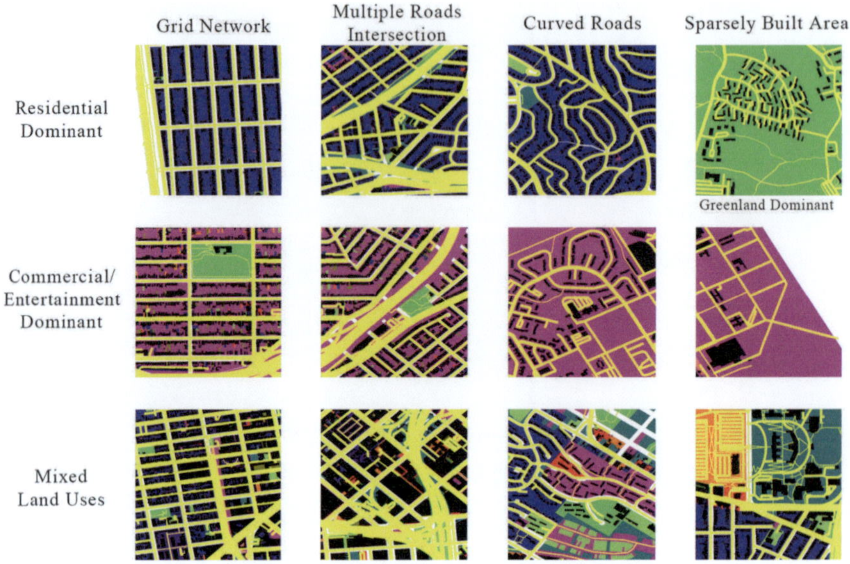

Fig. 13.9 The classification of the city plan

Our comparative analysis reveals that, for road patterns, areas with multiple road intersections have a significantly higher number of accidents surpassing grid networks, curved roads, and sparsely built area. Regardless of the land use type, areas with multiple road intersections experience the overall highest number of accidents, with incidents primarily concentrated at the intersections. Mixed land use areas, often characterized by higher road density, also report more accidents than single-use areas.

Overall, the layout with multiple road intersections is most prone to accidents. For grid networks, road density has a greater impact on accident numbers than land use type. Curved roads and sparsely built area generally show fewer accidents. Such analysis corresponds to the Marks study [16] that the distribution of crashes was fairly uniform across the gridirons and crash frequencies were dramatically higher for four-way than three-way intersections.

4.3 Urban Design Optimization

While theoretical research on the relationship between traffic accidents and urban road planning often remains abstract, our study aims to utilize big data and deep learning to offer urban planners more dynamic and actionable insights for design optimization.

We selected a district with high traffic accident density and complex land use patterns. Using Photoshop (PS), we modified the input base map colors by incorporating various forms of green spaces or other land use types into the existing site (see Fig. 13.10). These modified maps were then used as test inputs for the GAN model to predict traffic accidents, allowing us to assess whether different urban design strategies can effectively enhance traffic safety in the environment.

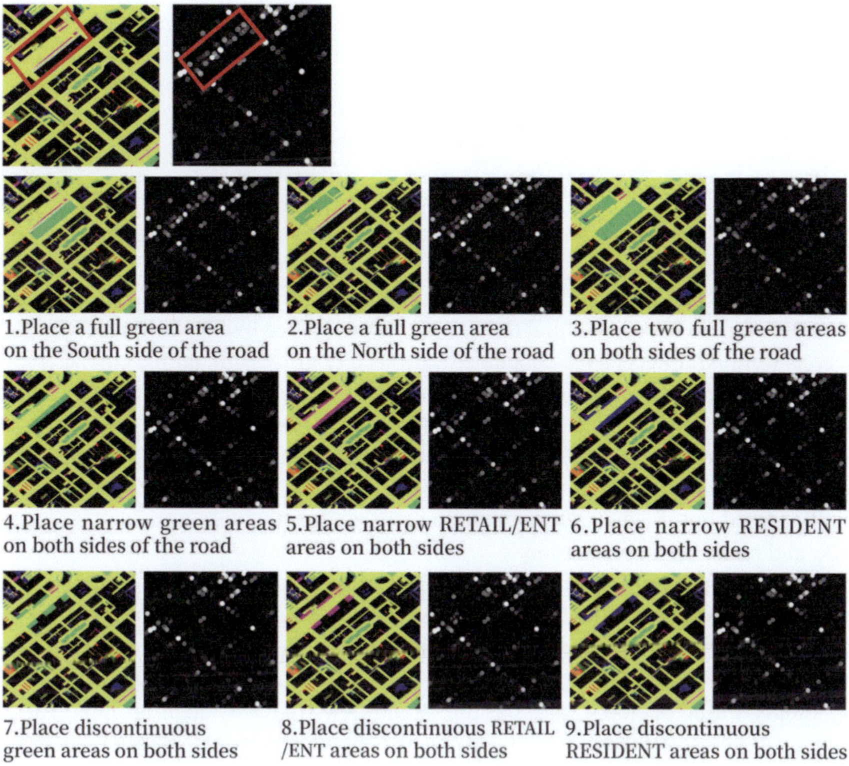

Fig. 13.10 The original site and its corresponding traffic condition

We conducted a multifaceted exploration of the relationship between built environment factors and traffic accidents from the perspective of designers:

- (Steps 1–4) Impact of Green Space Placement on Reducing Traffic Accidents:
  We added green spaces along the road on different sides and found that positioning green spaces on both sides is more effective in reducing traffic accidents. Given the practical constraints in urban planning that often prevent large-scale changes, we also explored the impact of reshaping green spaces into elongated forms. This adjustment proved more effective in reducing traffic accidents, indicating a larger green space does not necessarily lead to a greater reduction in the probability of traffic accidents. The effectiveness of green spaces in reducing traffic accidents is dependent on various factors, like the size, placement and shape.
- (Steps 5–6) Influence of Land Use Types on Traffic Accidents:
  We further investigated the effects of different land use types on traffic accident rates. Introducing narrow retail/entertainment areas or residential areas on both sides of the road impacts traffic accident reduction, but not significant among comparisons.
- (Steps 7–9) Effect of Spatial Segmentation on Reducing Traffic Accidents:

We examined the effect of spatial segmentation by dividing elongated green spaces. The results show that segmenting spaces can further reduce the occurrence of accidents, aligning with general urban design practices and highlighting the importance of segmentation in spatial planning.

4.4 Urban Design Theory Validation

Previous urban designs focus mostly on the pedestrians, but as transportation increasingly become part of people's life, there's a raising for need attention on traffic and perspective from drivers and passengers. Those theories raised by researches yesterday may not apply to the cities today. We want to find the city plans suitable for living while ensuring traffic safety as well. Architects have concluded various city grids with unique characteristics, which raises the doubt that if these grids applicable for some cities will apply to another one. Therefore, we chose five diverse cities (Barcelona, Paris, Manhattan, Sandiego and Suzhou) to test on the model specific trained for San Francisco. Among the five cities, Barcelona, Manhattan and Sandiego have obvious grids. Paris has a center with roads radially spread out. Suzhou seems more random, with little buildings share a common block, and roads surround outside.

As shown in Fig. 13.11, it turns out that city with gridions share a similar trend. Traffic accidents are more likely to happen on the wider roads. By contrast, the slim roads friendly to pedestrians and civilians according to Jane Jocabs [12], are welcoming to transportation as well. Though the practical urban design involves far more consideration than simply traffic accident rate, but this brand-new trial will open up mind for city designers to utilize the GAN model and experiment their imagined program on specific sites.

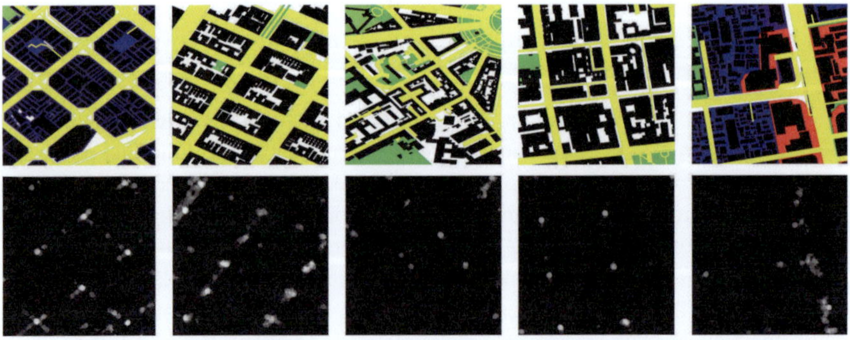

Barcelona Manhattan Paris Sandiego Suzhou

Fig. 13.11 The prediction of traffic accidents based on different city plans

4.5 Test on the Migration Capability of the Model

After the validation of the model in San Francisco, we dig deeper into the question: if the model can migrate to another city in America? Considering the climate and cultural similarities, we chose another two cities in western America-San Diego and Seattle. Both of the cities have gridions but with different density and land-use distribution. However, the model shows a low quality on the prediction of both cities' traffic accidents (see Fig. 13.12). While the model recognized the intersections and predicted a number of accidents, the dots are fewer than reality. Besides, the model fails to recognize the curved roads (especially highways) and presents no dots alongside these roads. In the future, we will further investigate on the prediction ability of our model and find its boundaries.

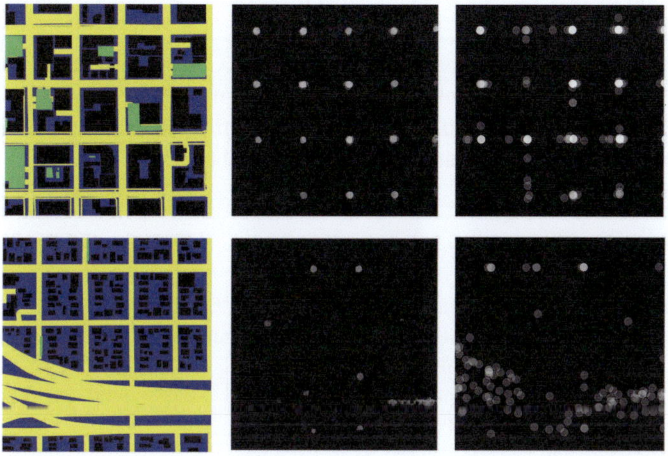

Fig. 13.12 The comparison between prediction and reality of traffic accidents in San Diego

5 Conclusion

5.1 Insights

Our study uses San Francisco as a case study to visualize urban built environment and traffic accident data through GIS and train a GAN model for traffic accident prediction.

Model Prediction: Our model has demonstrated high accuracy in predicting traffic accidents across various environments, including complex dense road networks, sparse road layouts dominated by green spaces, and areas with diverse land use types. This indicates that our model is capable of adapting to different urban layouts and traffic conditions in a specific city.

Validation of Urban Planning Theories: By analyzing and categorizing data based on road infrastructure and land use types, and comparing accident data, we have validated urban planning theories specific to the San Francisco region. Our simulation of different design scenarios further verifies and extends existing urban planning theories, providing a more scientific and reliable basis for urban planning decisions.

Designer-Friendly Urban Planning Visualization Method: We have introduced a dynamic prediction model to explore the impact of modifications to the built environment on traffic accidents. This approach offers a visualization tool tailored for designers, facilitating a better understanding of the implications of design changes. By applying neural network and big data advancements to urban planning, we provide a reference for interdisciplinary knowledge collaboration for architects and city designers.

5.2 Limitations and Future Work

Due to limitations in accessing urban traffic accident data, this model currently demonstrates good predictive accuracy only for San Francisco. However, this also suggests that traffic conditions may have unique characteristics across different cities, raising the question of whether to train a general model for urban prediction or to develop city-specific or region-specific models. This remains an area for further exploration. Future research should incorporate additional factors to analyze the causes of these differences beyond the built environment.

We plan to develop a web platform based on our GAN model, tailored to provide an intuitive interface for urban planners. This platform will enable designers to easily modify the built environment and receive real-time traffic accident prediction feedback, aligning with their design logic and reducing technical barriers.

References

1. Ahmed S, Mohammed M, Abdulqadir S, Abd Elkader R, El-Shall N, Chandran D, Ur Rehman ME, Dhama K (2023) Road traffic accidental injuries and deaths: a neglected global health issue. Health Sci Reports 6:e1240
2. Bergel-Hayat R, Debbarh M, Antoniou C, Yannis G (2013) Explaining the road accident risk: weather effects. Accid Anal Prev 60:456–465
3. Bremer, Stefanie, and Henrik Sander. "View from the Road: Environmental Simulation for the Fractal City of Rhine Ruhr." In Proceedings· 6th EAEA Conference· Bratislava, 43. 2003.
4. Chand A, Jayesh S, Bhasi AB (2021) Road traffic accidents: an overview of data sources, analysis techniques and contributing factors. Mater Today Proc 47:5135–5141
5. Chen Q, Song X, Yamada H, Shibasaki R (2016) Learning deep representation from big and heterogeneous data for traffic accident inference. In: Proceedings of the AAAI conference on artificial intelligence, vol 30, no 1
6. Ewing R, Dumbaugh E (2009) The built environment and traffic safety: a review of empirical evidence. J Plan Lit 23(4):347–367
7. Goniewicz K, Goniewicz M, Pawłowski W, Fiedor P (2016) Road accident rates: strategies and programmes for improving road traffic safety. Eur J Trauma Emerg Surg 42:433–438
8. Gutierrez-Osorio C, Pedraza C (2020) Modern data sources and techniques for analysis and forecast of road accidents: a review. J Traffic Transp Eng (English Edn) 7(4):432–446
9. Haddon W (1999) The changing approach to the epidemiology, prevention, and amelioration of trauma: the transition to approaches etiologically rather than descriptively based. Inj Prev 5(3):231–235
10. Heusel M, Ramsauer H, Unterthiner T, Nessler B, Hochreiter S (2017) Gans trained by a two time-scale update rule converge to a local nash equilibrium. Adv Neural Inf Process Syst 30
11. Imran M, Nasir JA (2015) Road traffic accidents: prediction in Pakistan. Professional Med J 22(6):705–709
12. Jacobs J (1961) The death and life of great American cities. Random House, New York
13. Kaygisiz Ö, Senbil M, Yildiz A (2017) Influence of urban built environment on traffic accidents: the case of Eskisehir (Turkey). Case Studies Transport Policy 5(2):306–313
14. Kusselson SB (2013) Investigating how land use patterns affect traffic accident rates near frontage road cross-sections: a case study on interstate 610 in Houston, Texas. Oklahoma State University

15. Liu Q, Chen C, Kuang X (2024) The nonlinear effect of land use patterns on the severity of pedestrian accidents—based on extreme gradient boosting decision tree approach. Sci Technol Eng
16. Marks H (1957) Subdividing for traffic safety. Traffic Quart, 308–25
17. McClure WR, Bartuska TJ (2011) The built environment: a collaborative inquiry into design and planning. Wiley
18. Mesquitela J, Elvas LB, Ferreira JC, Nunes L (2022) Data analytics process over road accidents data—a case study of Lisbon City. ISPRS Int J Geo Inf 11(2):143
19. Nguyen H, Kieu L-M, Wen T, Cai C (2018) Deep learning methods in transportation domain: a review. IET Intel Transport Syst 12(9):998–1004
20. Sobhan M (2023) US accidents (2016–2023). Kaggle. https://doi.org/10.34740/KAGGLE/DS/199387
21. Umair M, Rana IA, Lodhi RH (2022) The impact of urban design and the built environment on road traffic crashes: a case study of Rawalpindi, Pakistan. Case Studies Transp Policy 10(1):417–426
22. Vorko-Jović A, Kern J, Biloglav Z (2006) Risk factors in urban road traffic accidents. J Safety Res 37(1):93–98
23. Wang T-C, Liu M-Y, Zhu J-Y, Tao A, Kautz J, Catanzaro B (2018) High-resolution image synthesis and semantic manipulation with conditional gans. In: Proceedings of the IEEE conference on computer vision and pattern recognition, pp 8798–8807
24. Wang Y, Kockelman KM (2013) A poisson-lognormal conditional-autoregressive model for multivariate spatial analysis of pedestrian crash counts across neighborhoods. Accid Anal Prev 60:71–84
25. Yang BZ, Loo BPY (2016) Land use and traffic collisions: a link-attribute analysis using empirical bayes method. Accident Anal Preven 95:236–249
26. Zhang Z, Yang W, Wushour S (2020) Traffic accident prediction based on LSTM-GBRT model. J Control Sci Eng 2020(1):4206919
27. Zegeer CV, Council FM (1995) Safety relationships associated with cross-sectional roadway elements. Transp Res Record 1512
28. Chen J, Qiu Z, Peng T, Liu K, Fu Z, Tuo Y (2023) Influence of built environment on the severity of urban traffic accidents. J Chongqing Jiaotong Univ (Nat Sci) 42(3):105–111
29. AbelláN J, LóPez G, De OñA J (Nov 2013) Analysis of traffic accident severity using decision rules via decision trees. Expert Syst Appl 40(15):6047–6054
30. Muhammad LJ, Salisu S, Yakubu A, Malgwi YM, rufa'I Tijjani Abdullahi E, Mohammed IA, Muhammad NA (2017) Using decision tree data mining algorithm to predict causes of road traffic accidents, its prone locations and time along kano –wudil highway. Int J Database Theory Appl 10:197–206
31. Wu W, Jiang S, Liu R, Jin W, Ma C (2020) Economic development, demographic characteristics, road network and traffic accidents in zhongshan, china: gradient boosting decision tree model. Transportmetrica A: Transp Sci 16(3):359–387
32. Yuan Z, Zhou X, Yang T (2018) Hetero-convlstm: A deep learning approach to traffic accident prediction on het- erogeneous spatio-temporal data. In Proceedings of the 24th ACM SIGKDD international conference on knowledge discovery & data mining, pp 984–992
33. Yu H, Wu Z, Wang S, Wang Y, Ma X (2017) Spatiotemporal recurrent convolutional networks for traffic prediction in transportation networks. Sensors 17(7):1501
34. Zhang Y, Li Y, Zhou X, Kong X, Luo J (2020) Curb-gan: Conditional urban traffic estimation through spatio- temporal generative adversarial networks. In Proceedings of the 26th ACM SIGKDD International Conference on Knowledge Discovery & Data Mining, pp 842–852
35. Huang J, Cui Y, Zhang L, Tong W, Shi Y, Liu Z (2022) An overview of agent-based models for transport simulation and analysis. J Adv Transp

36. Wang J, Lv W, Jiang Y, Qin S, Li J (2021) A multi-agent based cellular automata model for intersection traffic control simulation. Physica A-Stat Mech Appl 584:126356
37. Abdulsattar H, Siam MRK, Wang H (2020) Characterisation of the impacts of autonomous driving on highway capacity in a mixed traffic environment: an agent-based approach. IET Intell Transp Syst
38. Torabi B, Wenkstern RZ, Saylor RA (2018) Collaborative agent-based traffic signal system for highly dynamic traffic conditions. In 2018 21st International Conference on Intelligent Transportation Systems (ITSC), pp 626–633
39. Zhang Y, Li Y, Zhou X, Kong X, Luo J (2019) Trafficgan: Off-deployment traffic estimation with traffic generative adversarial networks. In 2019 IEEE international conference on data mining (ICDM), IEEE, pp 1474–1479
40. Sun S, Mu L, Feng R, Wang L, He, J (2021) Gan-based lucc prediction via the combination of prior city planning information and land-use probability. IEEE J Sel Top Appl Earth Obs Remote Sens 14:10189–10198
41. Nguyen K-T, Dinh D-T, Do MN, Tran M-T (2020). Anomaly detection in traffic surveillance videos with gan- based future frame prediction. In Proceedings of the 2020 International Conference on Multimedia Retrieval, pp 457–463
42. Chen Z, Zhang J, Zhang Y, Huang Z (2021) Traffic accident data generation based on improved generative adversarial networks. Sensors 21(17):5767
43. Jin J, Rong D, Zhang T, Ji Q, Guo H, Lv Y, Ma X, Wang F-Y (2022)A gan-based short-term link traffic prediction approach for urban road networks under a parallel learning framework. IEEE Trans Intell Transp Syst 23(9):16185–16196
44. Isola P, Zhu J-Y, Zhou T, Efros AA (2017) Image-to-image translation with conditional adversarial networks. CVPR
45. Huang W, Zheng H (2018) Architectural drawings recognition and generation through machine learning. In Proceedings of the 38th annual conference of the association for computer aided design in architecture, Mexico City, Mexico, pp 18–20
46. Liu Y, Zhang Z, Hu K, Deng Q (2023) Graph constrained multiple schemes generation for campus layout. In The International Conference on Computational Design and Robotic Fabrication, Springer, pp 125–138
47. He J, Zheng H (2021) Prediction of crime rate in urban neighborhoods based on machine learning. Eng Appl Artif Intell 106:104460
48. Quan SJ (2022) Urban-gan: An artificial intelligence-aided computation system for plural urban design. Environ Plan B: Urban Anal City Sci 49(9):2500–2515

Digital Analysis and Design for Historical and Cultural Districts from the Perspective of Spatiotemporal Vitality: Using Hangzhou Qiaoxi District as the Case

Ziyi Chen and Yuxuan Liu(✉)

College of Architecture and Urban Planning, Tongji University, Shanghai, China
{sherrychen_czy,yuxuanliu}@tongji.edu.cn

Abstract. The protection and renewal of historical and cultural districts are receiving increasing attention. Nevertheless, many of them undergo huge gaps in vitality between day and night and existed researches lack consideration of vitality assessment in multiple time dimensions. As a response to this situation, based on multi-source data and new technologies, we proposed a digital analysis and design method for historical and cultural district from the perspective of spatiotemporal vitality. Key evaluation dimensions were extracted and analyzed by integrating classic theory of urban design and spatial vitality, which were then quantitatively analyzed and illustrated to discover areas with insufficient spatiotemporal vitality, thus allowing for targeted strategies and design. Moreover, utilizing the quantitative benefits provided by multi-source data, the continuation of area's spatiotemporal vitality and the enhancement of its spatial richness can be quantitatively assessed. In short, this study contributes to an exploration of the relationship between the spatial vitality, functional characteristics and spatial perception of historical and cultural district under spatiotemporal dynamics. And it establishes a method for quantitatively evaluating design effectiveness, while conducting design intervention and effectiveness evaluation. Meanwhile, it makes a step forward for providing embedded support across multiple stages from status analysis to design and evaluation, so that provides method for the consistency of spatiotemporal vitality in historical and cultural districts.

Keywords: Historic and cultural district · Spatiotemporal vitality · Digital analysis · Urban design · Multi-source data · Design science

1 Introduction

Historical and cultural district, a vital component of China's urban architectural heritage [1], whose protection and renewal has become a critical issue in urban planning and cultural construction [2]. While many historical and cultural districts have been successfully preserved through practical efforts in recent years [3], however, some experience stark disparities in day-and-night vitality: Pedestrian flow drops sharply and leading to a "cliff-like" decline in activity after evening hours. Since consistent vitality is crucial

J. Fu et al. (eds.), *AI for Architecture*, Computational Design and Robotic Fabrication,
https://doi.org/10.1007/978-981-95-0974-4_14

for the sustainable development of historical and cultural districts [4], characterizing its mechanisms and enhancing continuity are of paramount importance.

Meanwhile, the emergence of analytical techniques and urban data bring new research potentials to the innovation of precise digital analysis and design methods [5], shifting urban analysis and evaluation away from mere reliance on manual method and subjective judgment [6]. As an important part of urban space, historical and cultural districts can be provided a foundation for precise analysis and guiding updates for sustained spatiotemporal vitality by thoroughly analyze the element characteristics of spatial vitality across multiple time dimensions based on multi-source data and new technologies.

2 Literature Review

2.1 Current Spatial Evaluation Methods for Historical and Cultural District

In the early stages, the spatial evaluation of historical and cultural districts is primarily guided by classical urban space theories and relied heavily on subjective and manual research methods, focusing on cultural value and user satisfaction. For instance, semi-structured surveys combined with interviews are used to investigate the relationship between place attachment and shopping satisfaction in historical districts [7], or to assess the modern adaptability of these districts through public and expert surveys [8]. These methods reveal the spatial characteristics and qualities of historical and cultural district to some certain extents, however, due to the over-reliance questionnaire and on-site observation, which inevitably introduce subjective bias from a small sample and fail to accurately capture dynamic changes in spatial vitality.

Meanwhile, the raising of multi-data and new technologies has brought the widespread use of quantitative methods in the spatial evaluation of historical and cultural districts. For instance, the concentration of people in district can be accurately describe by collecting and analyzing mobile phone signaling and social media data [9], or urban renewal can be evaluated using the fuzzy analytic hierarchy process (AHP) [10]. However, these quantitative evaluation methods don't offer accurate spatial vitality analysis from a spatiotemporal dynamic perspective.

Recently, spatial evaluation and design under spatiotemporal dynamics start to gain attention, e.g., Territorial Timetable Plan (TTP), which improves facility utilization efficiency by coordinating opening hours and personal time and space accessibility [11], "Space–Time-People", a multi-dimensional cognitive model, assesses the dynamic integrity of historical landscapes [12]. This approach has also been used in transportation planning, where pedestrian and vehicle flow data in historical districts are collected for space syntax analysis [13]. These methods provide new insights into the temporal characteristics of spatial vitality in historical districts. However, they have not yet established a systematic analytical dimension, making it difficult to fully reflect the true state of neighborhood vitality. Additionally, many of them overlook spatial perception at the human scale, with little design interventions or effectiveness evaluations based on the analysis results.

2.2 Key Dimensions of Accurate Analysis of Historical and Cultural Districts

This study focuses on social vitality, defined as the degree of congregation of people participating in social activities in a certain area. It includes two aspects: one is the various activities that people can engage in, the other is the spaces supporting these activities. The former emphasizes human-centered spatial perception, such as Kevin Lynch's concept of "sensory quality" [14], the latter focuses on the impact of functional and morphological features in the environment on vitality, as highlighted by Jane Jacobs' assertion of the importance of functional diversity for the sustainability and vibrancy of a district [2]. Therefore, this study systematically analyze the spatial vitality of historical and cultural districts from two perspectives: functional characteristics and spatial perception.

Additionally, to achieve precise digital analysis on a systematic basis, four key dimensions are selected based on classic urban design theory. These factors not only reflect the functions and spatial perception characteristics of historical districts, but are also measurable using existing techniques. That is, *functional diversity, supply–demand coupling, accessibility, and spatial permeability* (Table 14.1), serve as the main entry points for precise digital analysis and design intervention.

Table 14.1 Key dimensions of digital analysis of historical and cultural district from spatiotemporal vitality perspective [15–19]

Proposers dimensions	Jane Jacobs	Kevin Lynch	William H. Whyte	Jan Gehl	Ashihara Yoshinobu
Functional diversity	Diversity is the condition of city life	Emphasize consistent morphological characteristics and dominant functions	Various activities should be concentrated and mixed	Different people, different groups, and in different situations have different requirements and different tolerances for space	Streets are not just for traffic; they exist as part of the community
Supply–demand coupling	The trick is to disperse time and use so that no part of the day is too empty or too full	The city should accommodate people of different classes and personalities	Supply creates demand; there are differences in place preferences among different groups of people	A good city is built on the body and senses of people	Buildings and streets are based on human nature
Accessibility	Communities should be the building blocks of cities	Focus on the directional accessibility and systematization of streets	Social gathering places are located at or very near intersections of traffic routes and plazas	Urban design should prioritize the needs of pedestrians	/
Spatial permeability	Short blocks and irregular streets, with many corners and changes in direction, provide diversity and richness, and create a sense of surprise	/	(People sitting in plazas) are seen on the horizontal plane of vision…	Winding or varied streets make walking more interesting	External space is a building without a roof

2.3 Emergence of Analysis and Design Methods: Multi-dimensional and Multi-link Support

The emergence of multi-source data and new technologies provides support for accurate digital analysis and design in urban space, whose development process under the four dimensions obtained in 1.2 is shown in Fig. 14.1. Early researches primarily focus on analyzing situation and identifying problem through a single-dimensional approach. For instance, studies quantifying functional diversity often extract activity characteristics and population density using mobile phone signaling or POI data [20–22]. In terms of supply–demand coupling, besides research on traffic issues in historical districts [23, 24], spatial vitality is often measured by analyzing POI data to represent supply (through service radius and density) and social media data to reflect demand [1, 25, 26]. And many studies have used methods such as space syntax and spatial network analysis technology to measure accessibility [27–29]. In terms of spatial permeability, direct methods base on space syntax while indirect ones use AHP to measure people's perception [10].

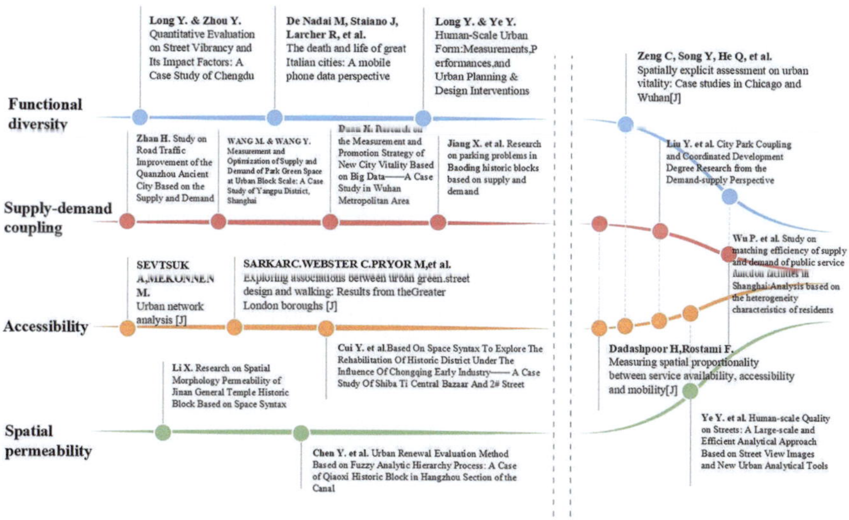

Fig. 14.1 Research development of four analysis dimensions based on multi-source data and new technologies

With the development of precision analysis researches, there has been a shift from single-dimensional to multi-dimensional analysis. For instance, evaluating spatial vitality based on four aspects (population density, livability, accessibility and diversity) [30], measuring spatial fairness based on three factors (service availability, accessibility and mobility) [31], analysis of street quality and accessibility using SVI and machine learning [32], and comprehensive analyses of facility supply using survey and POI data [33]. All of them reflect the trend of collaborative use of multi-source data and new technologies for comprehensive spatial evaluation, facilitating efficient and accurate digital analysis of urban spaces.

However, current comprehensive assessments still have limitations: (1) There is litter comprehensive and accurate method for analyzing the spatial vitality of historical and cultural districts across spatiotemporal dimensions. (2) A multi-step quantitative approach that integrates status quo analysis, problem diagnosis, design intervention, and effectiveness evaluation has yet to be developed. Therefore, exploring a set of digital analysis and design intervention methods from the perspective of spatiotemporal vitality to provide multi-dimensional and multi-link support for the oriented renewal of historical and cultural districts becomes a pressing need.

3 Method

3.1 Research Framework

This study contains four main phases. First, a theoretical framework was developed by combining multi-dimensional and multi-step analysis to address current issues in historical districts and the limitations of existing research. Second, four key dimensions—*functional diversity, supply–demand coupling, accessibility, and spatial permeability*—were analyzed in the context of the *time* factor using advanced urban analysis technologies. Third, case analysis and design were carried out by importing data of certain historical district to analyze its current spatiotemporal vitality and propose improvement strategies. Finally, the renovation plan was quantitatively evaluated using the method developed in the first phase. In short, this study created a digital analysis and design workflow for historical and cultural districts, utilizing data as the foundation, quantification as the method, and accurate design as the goal, all from the perspective of spatiotemporal vitality (Fig. 14.2).

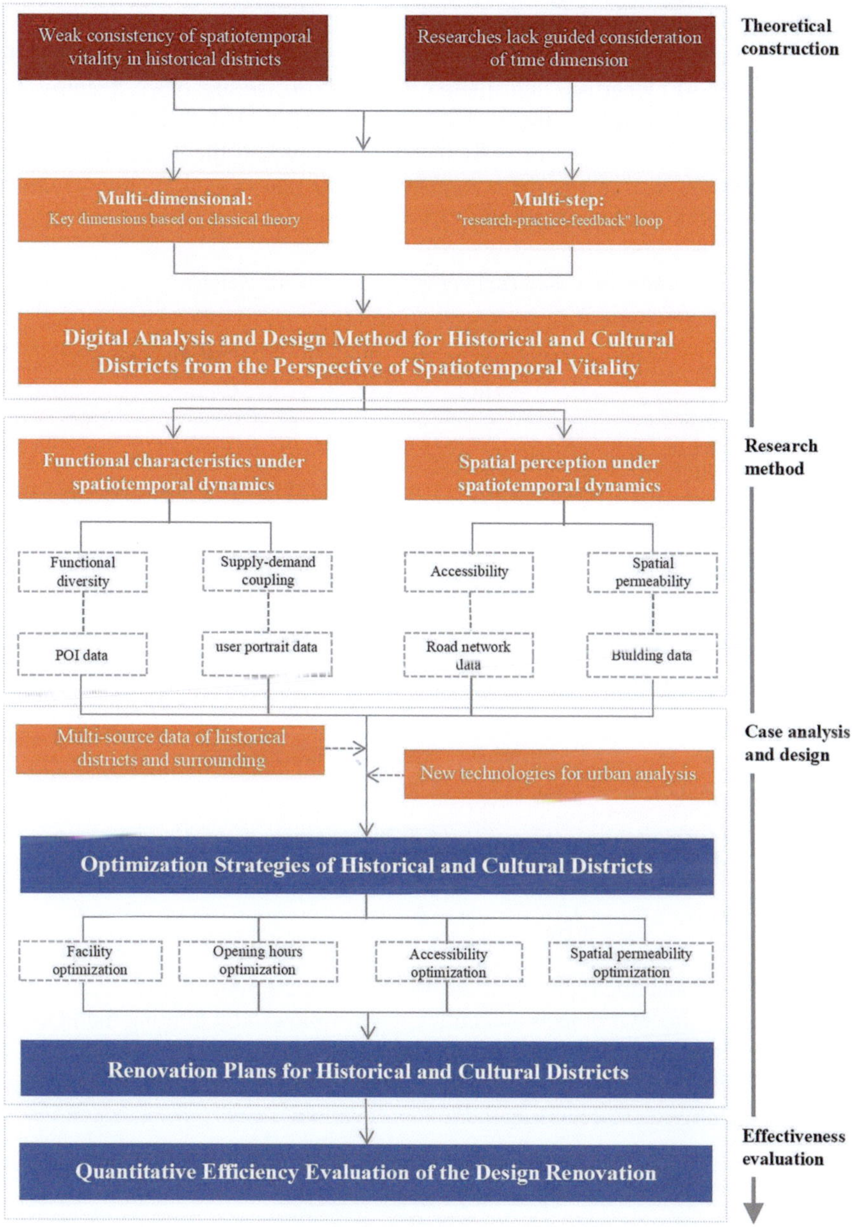

Fig. 14.2 Research framework

3.2 Research Method: Technical Path for Accurate Measurement of Analytical Dimensions

In the data preparation stage, the types of facilities to be analyzed and their corresponding weights were first determined. Here in, the AHP was used to calculate the impact weight of six facility types which were the main components of historical districts (CR: 0.069 < 0.1, passed the consistency test, as shown in Table 14.2). Additionally, POI data retrieved via an API interface and filtered according to facility types. Finally, AHP was applied again to calculate the weights of the four key analytical dimensions that influenced spatiotemporal vitality (CR: 0.059 < 0.1, passed the consistency test, as shown in Table 14.3), enabling the comprehensive spatiotemporal vitality results and corresponding vitality diagrams for the region.

Table 14.2 Key facilities and corresponding weights

Facility type	Eigenvector	Weight (%)	Max eigenvalue	CI
Catering services	1.976	32.93	6.437	0.087
Educational and cultural facilities	1.672	27.86		
Entertainment facilities	0.944	15.73		
Transportation services	0.609	10.16		
Commercial facilities	0.588	9.80		
Healthcare facilities	0.211	3.52		

Table 14.3 Key analysis dimensions and corresponding weights

Dimensions	Eigenvector	Weight (%)	Max eigenvalue	CI
Functional diversity	1.956	48.9	4.156	0.052
Supply–demand coupling	1.232	30.79		
Accessibility	0.612	15.31		
Spatial permeability	0.2	5		

In the term of functional diversity, the 15-min walk-catchment area was simulated to define the experimental objects and buffer zones, followed by filtering the POI data. The Shannon Index, combined with the facility weights, was then used to calculate the diversity of facilities in the district during both morning and evening periods. Specifically, it was subsequently calculated using the follow equation:

$$H = -\sum_{i=1}^{R} w_i * p_i ln(p_i) \tag{14.1}$$

where H is the POI diversity index and R is the total number of POIs, p_i is the relative abundance of the POI, that is, the proportion of a certain type of POI among all POIs in the grid, w_i is the weight coefficient.

In terms of supply–demand coupling, the first step was to measure the differences between supply and demand for six key facility types. Specifically, data on facility status and user profiles were collected through map platform data and surveys. Taking the opening hours and the number of facilities as the supply side, and the visiting time period and the number of people for each facility as the demand side, the supply and demand overlap was calculated and weighted by the number of users in each time period to determine the supply–demand coupling degree. Secondly, to identify areas for design intervention, the facility type with the lowest coupling was selected for further classification. A refined supply–demand coupling analysis was then conducted for these facilities. The specific formula is as follows:

$$C_n = \sum_{i=1}^{6} \frac{T_s \cap T_d}{T_d} \times w_i \tag{14.2}$$

$$CD_n = \frac{C_1 + C_2 + \cdots + C_n}{n} \tag{14.3}$$

where CD_n is the supply–demand coupling of a certain facility type while C_n represents supply–demand coupling of a certain facility. And T_s is the open time of a space while T_d is users' main activity time period (24 h is divided into 6 equal segments) and w_i is the weight of prefer using for each type.

In term of accessibility, *choice* reflected the frequency of a road being the "primary route" connecting other roads based on space syntax, thus representing the accessibility of the road network.

In terms of spatial permeability, visibility graph analysis (VGA) was used to examine the hierarchical relationship of spaces, including the separation, interspersion, opening up and connection. And *connectivity* indicated the number of other elements visible from a given element, thereby characterizing spatial permeability.

Finally, based on the precise analysis and problem judgment of the key dimensions mentioned above, the region's comprehensive spatiotemporal vitality was calculated, followed by the proposal of renewal strategies, design intervention, and performance evaluation. The effectiveness of the transformation was then assessed by comparing the improvements in spatiotemporal vitality and consistency before and after the intervention.

4 Results: Current Status of Hangzhou Qiaoxi Historical and Cultural District Based on Quantitative Analysis

4.1 Study Area

Our analysis was conducted in Hangzhou Qiaoxi Historical and Cultural District (hereinafter—Qiaoxi), a historical reserve in the center of the city integrating the daily life of citizens with the distinctive industrial heritage (Fig. 14.3). Considerable museums renovated from factory buildings, which had fostered the integrated development of exhibitions, commerce and parks, making it an important part of local cultural tourism. However, there was still a problem of insufficient nighttime vitality and gaps in spatiotemporal vitality. Given its geographical characteristics and availability of multi-source data, Qiaoxi had the operability for digital analysis and design intervention, making it the study object.

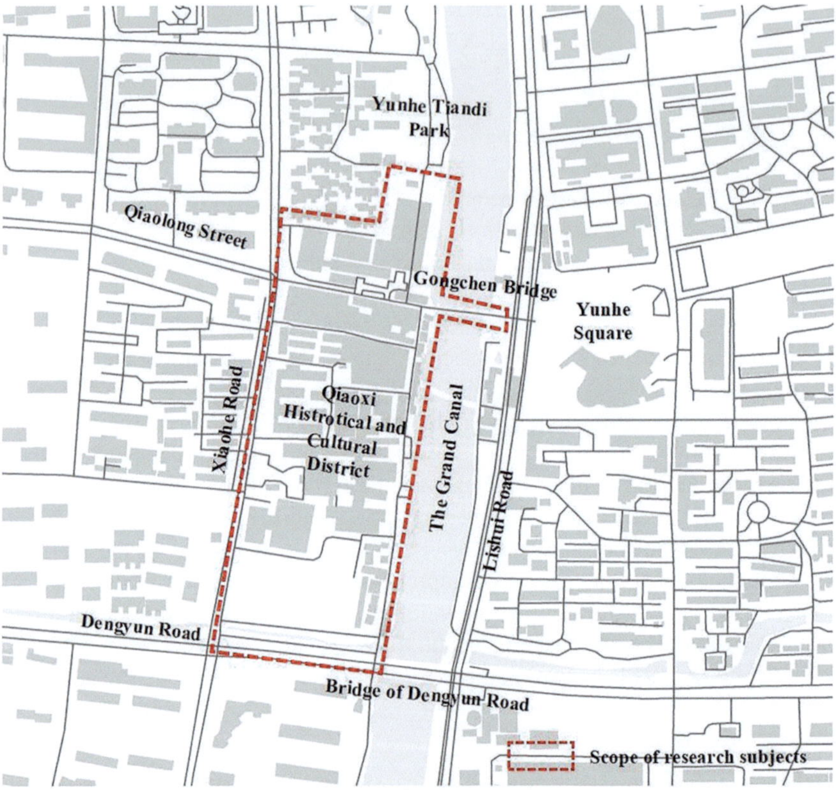

Fig. 14.3 Study area: Qiaoxi Historical and cultural district

4.2 Functional Diversity Based on POI Data

Figure 14.4 shows the functional diversity of Qiaoxi both during day and night, highlighting the differences across various times and identifying existing problems. POI data was crawled via the open map platform API and filtered based on a 15-min walking radius. According to the spatial scale characteristics of the historical district, an analysis fishnet was established based on 50 m × 50 m, and using a 100 m radius for analysis and a near table was generated. The functional diversity of spatial units were analyzed based on weights of different facility types (Shown in Table 14.1).

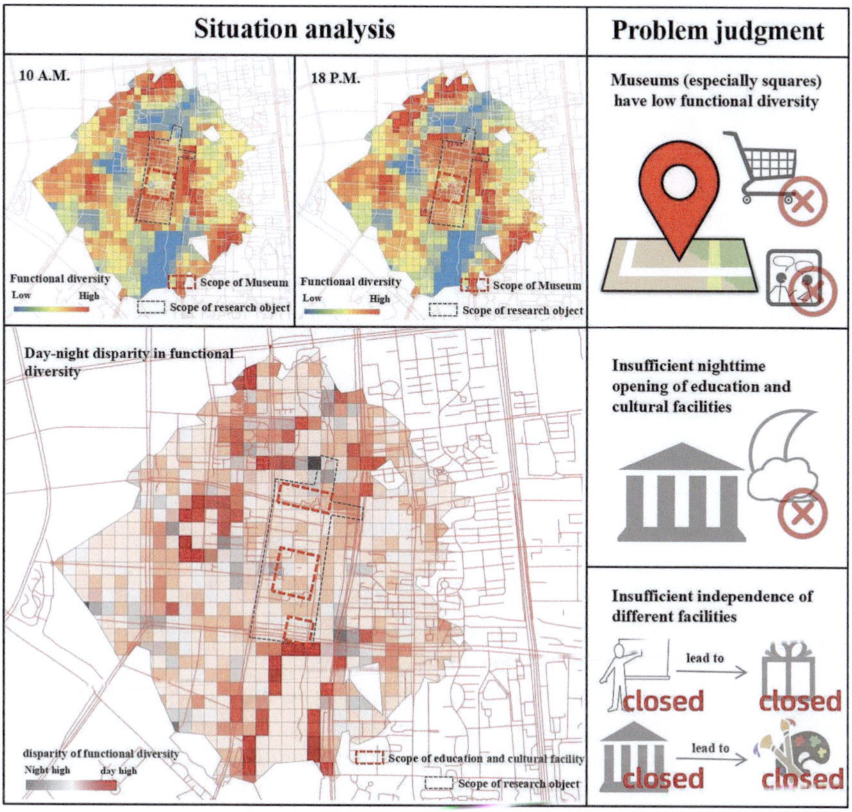

Fig. 14.4 Analysis result of functional diversity based on POI data

As shown in diagrams, although the functional diversity of Qiaoxi is relatively high at the same time, aligning with its role as a key cultural tourism, the diversity in the central section sharply drops compared to the surrounding area. Comparing under dynamic time dimensions, functional diversity is higher during the day than at night, especially around the Chinese Fan Museum, Chinese Knife and Scissors Museum, as well as the pedestrian streets to the east. These areas, predominantly consisting of educational and cultural facilities, close after 18:00, causing other facilities shared entrances to close as well, reducing the functional diversity.

4.3 Supply–Demand Coupling Based on Amap and User Portrait Data

Figure 14.5 shows the supply–demand coupling of Qiaoxi. First, 191 valid questionnaires were collected to record the users' preferences for different types of 6 main facilities mentioned in Sect. 3.2, creating a demand-side 'time-function' mapping. Then, the importance weight was calculated based on the number of people preferring each facility type, and the opening hours were compared with the visiting time and demand to calculate the supply–demand coupling.

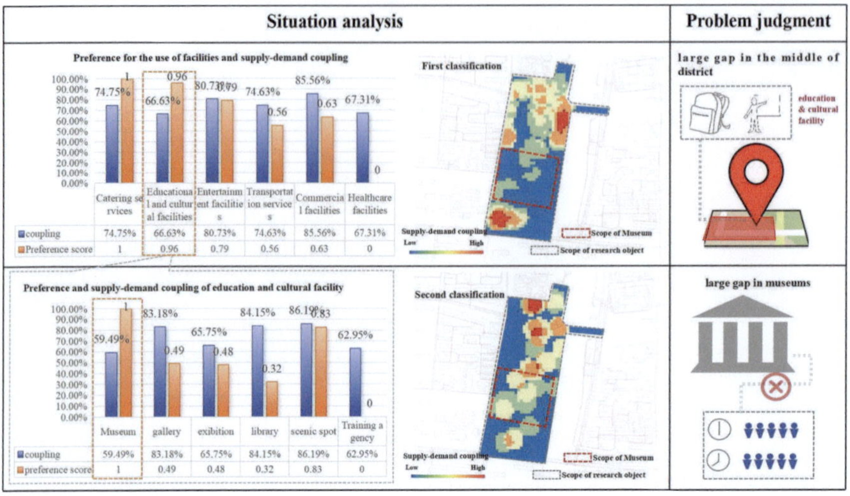

Fig. 14.5 Analysis result of supply–demand coupling based on Amap and user portrait

The results show that more users tend to visit Qiaoxi in the evening, with a preference for educational and cultural facilities. However, these facilities have the lowest supply–demand coupling, highlighting the need for focused design and operational strategies, especially for nighttime use. Based on this preliminary result, the supply–demand coupling for subcategories of educational and cultural facilities was further analyzed to identify core areas and functions for design intervention. Here in, educational and cultural facility is divided into six categories: museums, art galleries, exhibition halls, libraries, scenic spots and training institutions, calculating the supply–demand coupling through the same method above.

As the Fig. 14.5 shows, museums have the largest spatial preference but the lowest supply–demand coupling, indicating a significant gap between supply and demand, particularly after 16:00. Targeted strategies are needed to address the demand for nighttime museum visits.

4.4 Accessibility Based on Space Syntax

Figure 14.6 shows the road accessibility of Qiaoxi both during day and night, showing road choice calculations, normalization processing, and a comparison of differences. This study used the 15-min living circle as the buffer zone to establish road network, and used Depthmap Beta1.0 to calculate the accessibility by analyzing the frequency with which a space appears on the shortest topological path.

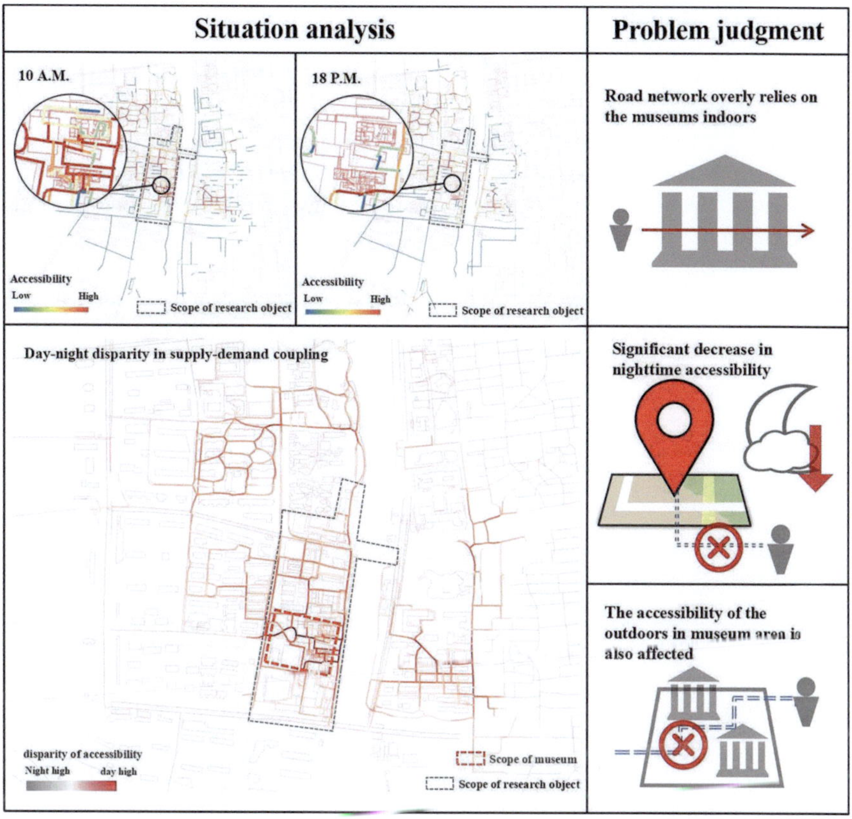

Fig. 14.6 Analysis result of accessibility based on space syntax

As shown in diagrams, within the same time dimension, the road network is relatively complex and the accessibility is generally high, creating a rich spatial experience with intertwined streets. However, a large number of roads in the museum area need to pass through internal spaces, making accessibility highly dependent on the museum's opening hours. When comparing under dynamic time dimensions, accessibility is much higher during the day than at night, particularly around the museums and surrounding squares. Additionally, although there are roads connecting different squares, many of them are narrow without lighting guidance at night, leading to a large spatiotemporal difference after the museum is closed.

4.5 Spatial Permeability Based on VGA

Figure 14.7 shows the spatial permeability of Qiaoxi both during day and night, not only showcasing the permeability of streets and squares but also incorporating the internal spaces of public buildings into the analysis. To comprehensively simulate people's spatial experience, floor plans including walls, doors and windows were drawn, followed by VGA analysis. Spatial connectivity was then calculated, normalized, and differences were compared.

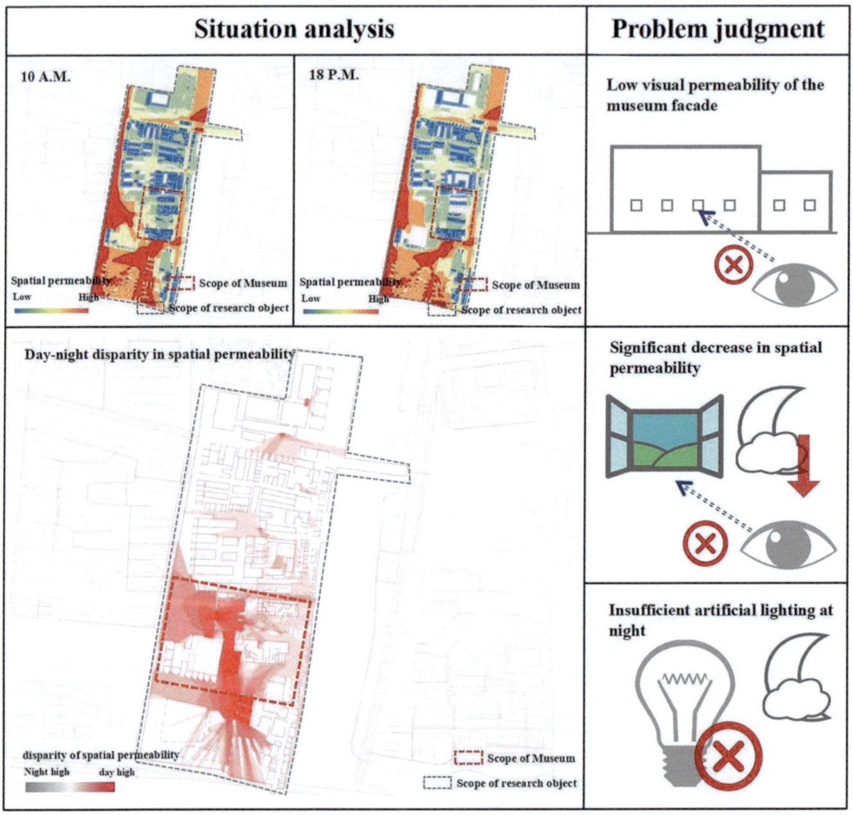

Fig. 14.7 Analysis result of spatial permeability based VGA

As shown in diagrams, within the same time dimension, the spatial permeability of the museum area is relatively low. It is because most of the museums are located in repurposed factory buildings, where the façades have retained the original design, with small windows limiting visual permeability. When comparing under dynamic time dimensions, the permeability is much lower during the night than at day, especially around museums for two main reasons. First, a large number of educational and cultural facilities are closed after evening, making it difficult to achieve spatial penetration in the entrance. Second, as natural light decreases, the main building and surrounding spaces fail to provide enough artificial light, resulting in reduced spatial permeability at night.

4.6 Overlay Analysis of Spatiotemporal Vitality Based on AHP

Figure 14.8 shows the comprehensive spatiotemporal vitality of Qiaoxi both during day and night, overlaying the importance weight of the four key analysis dimensions in Sect. 2.2. Within the same time dimension, the spatiotemporal vitality of the northern, central and southeastern areas is reletively low with the main functions of educational and cultural facilities. Notably, the spatiotemporal vitality of the central museum cluster is disconnected from the north and south areas. When comparing under dynamic time dimensions, the consistency of vitality in Qiaoxi is weak, eparticularly in the museum cluster area in the central and southeastern parts. Therefore, strategies need to be proposed for this area to improve the consistency of spatiotemporal vitality.

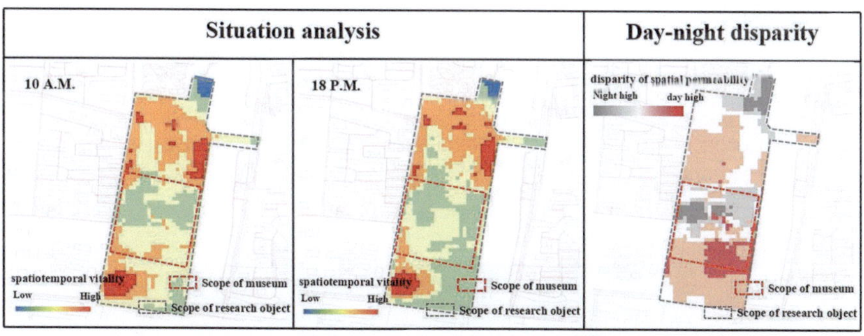

Fig. 14.8 Analysis result of comprehensive spatiotemporal vitality in Qiaoxi

5 Strategic Response and Effectiveness Evaluation

5.1 Strategic Response

Through digital analysis, 4 aspects and 9 types of issues of spatiotemporal vitality issues in Qiaoxi have been identified. Specifically, the core issue lies within educational and cultural facilities, particularly the museum cluster and its surrounding public spaces.

These areas are large and attractive but typically close after 16:30 and have not been adapted for the nighttime environment, resulting in a decline in vitality and impacting the entire district. Currently, the spatial perception of these areas is greatly affected by whether the museums are open, however, accessibility and perceptibility are not identical. Therefore, targeted strategies can be used to enhance the vitality of the space throughout the entire day without altering their operating hours. Four strategies will be discussed in the following section to guide design intervention, followed by an evaluation of their effectiveness (Fig. 14.9).

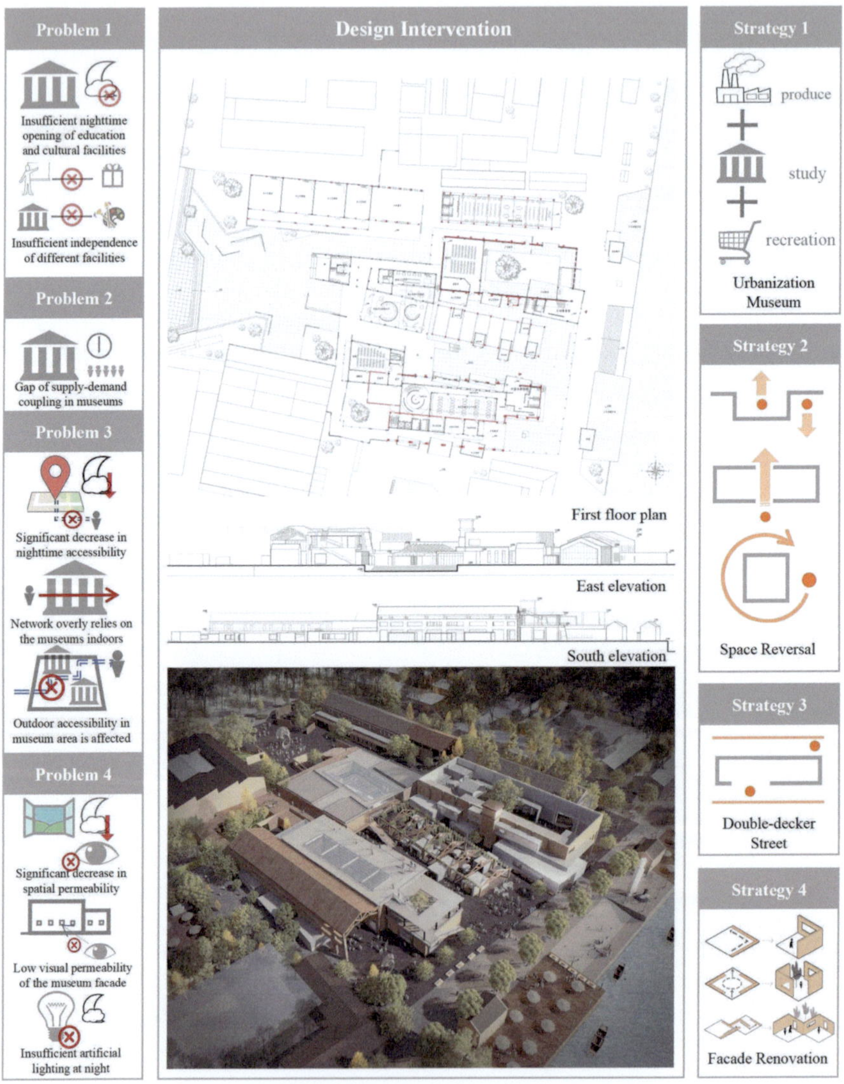

Fig. 14.9 Diagram of issue summary and strategy response

Promotion Strategy Response of Functional Diversity

Figure 14.10 shows the promotion strategy for functional diversity and a specific functional optimization plan, called as the concept of "urbanization museum", encompasses two main ideas. First, the museumization of city: combined with the historical background of Qiaoxi's handicrafts, industry and canal commerce, exhibitions were integrated into public activities and district itself was turned into a museum. Second, the urbanization of museums: exhibition expanded beyond traditional showcase displays, becoming a more flexible and dispersed space that interwove with creative business, handicraft display, etc.

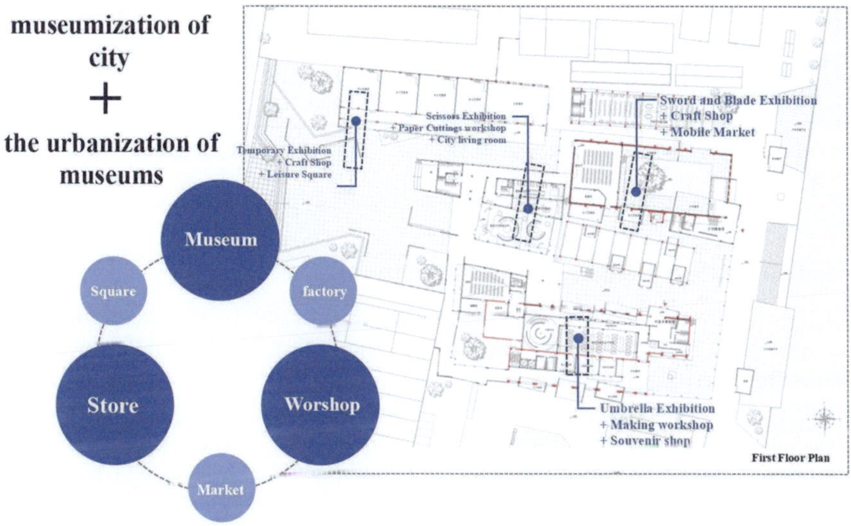

Fig. 14.10 Design drawing of "Urbanization Museum" in Qiaoxi

Promotion Strategy Response of Supply–Demand Coupling

Figure 14.11 shows the promotion strategy for supply–demand coupling, referred to as "space reversal", that is, the same space can be perceived from both inside and outside at different times of the day, enabling visibility of exhibits at night without changing the museum's opening hours. Specifically, it included three approaches: (1) Reversal of walls, where zigzag walls created independent but visually connected spaces through window openings; (2) Reversal of streets: where thickened walls integrated functional spaces and increased the visual transparency to reveal the exhibitions behind; (3) Reversal of buildings: for the exhibition with strong integrity, embedding transparent display cabinets in facades to showcased exhibits from the street outside the building, even when interiors were closed at night. Overall, these methods not only ensured continuous exhibit viewing, but also enriched the viewing and space experience.

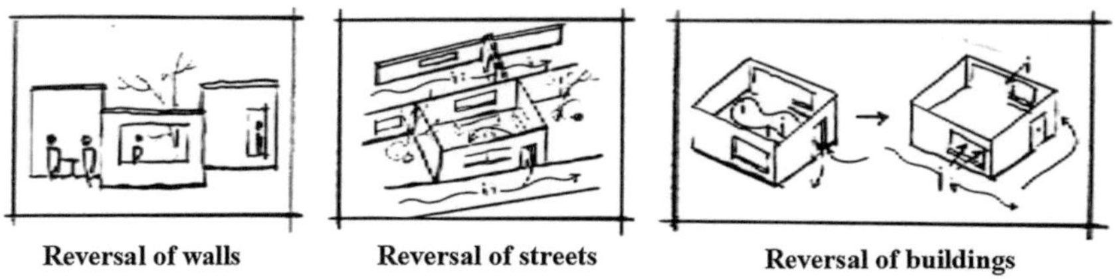

Fig. 14.11 Draft of "Space reversal" in Qiaoxi

Promotion Strategy Response of Accessibility

Figure 14.12 shows the "double-decker street" strategy for improving accessibility, based on a typological study of the street texture in Qiaoxi, that is, two parallel streets with different characteristics positioned in front of and behind functional spaces. These street types were then applied to the renovation design. Specifically, the existing exhibition interface was shifted inward a few meters to form two parallel routes indoors and under the semi-outdoor gallery, which could not only extend the historical street texture into the museum area, but also enabled independent road management, increasing the nighttime accessibility of the exhibition spaces.

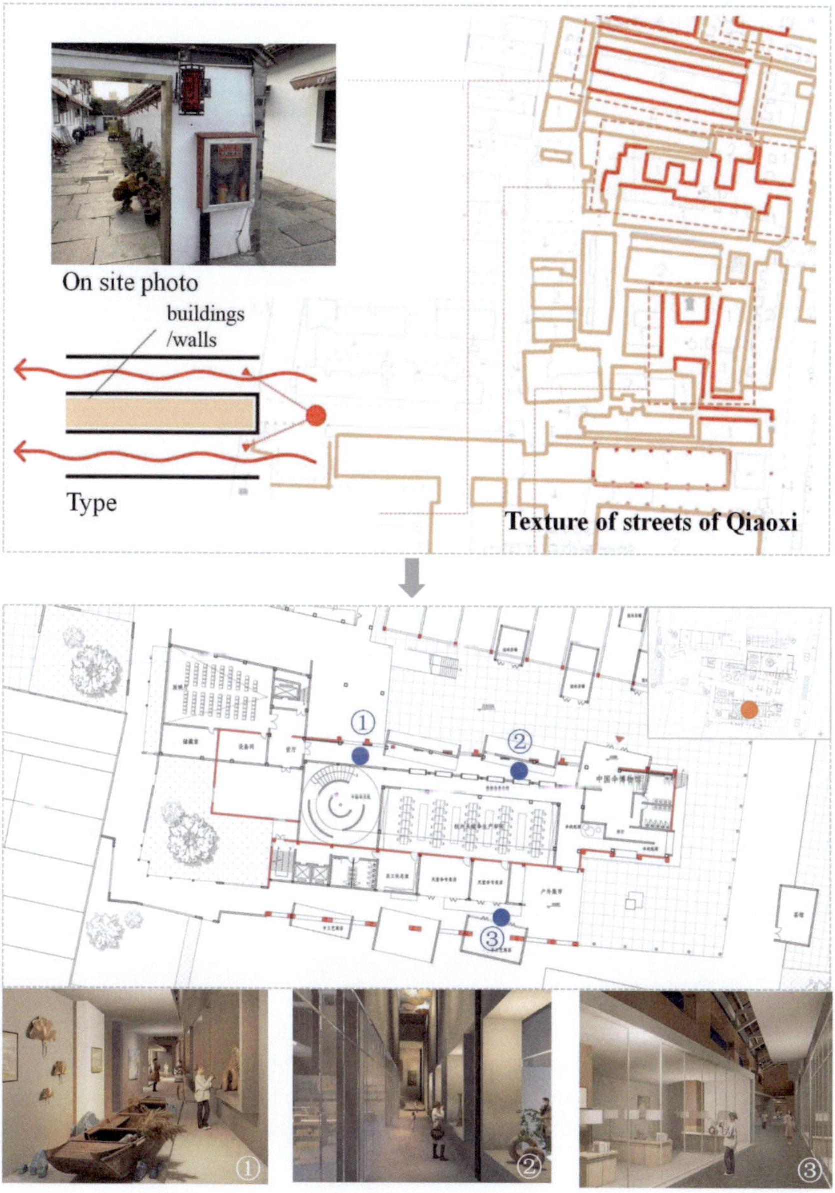

Fig. 14.12 Diagram of "Double-decker street" in Qiaoxi

Promotion Strategy Response of Spatial Permeability

Figure 14.13 shows the "facade renovation" strategy for spatial permeability, informed by human-centered perception and visual analysis. Specifically, through the typological study of the wall system and the the behavioral relationship between the wall and people, especially the location and size of the windows, facade renovation were carried out. This approach enhanced spatial permeability while creating richer spatial variations and visual effects.

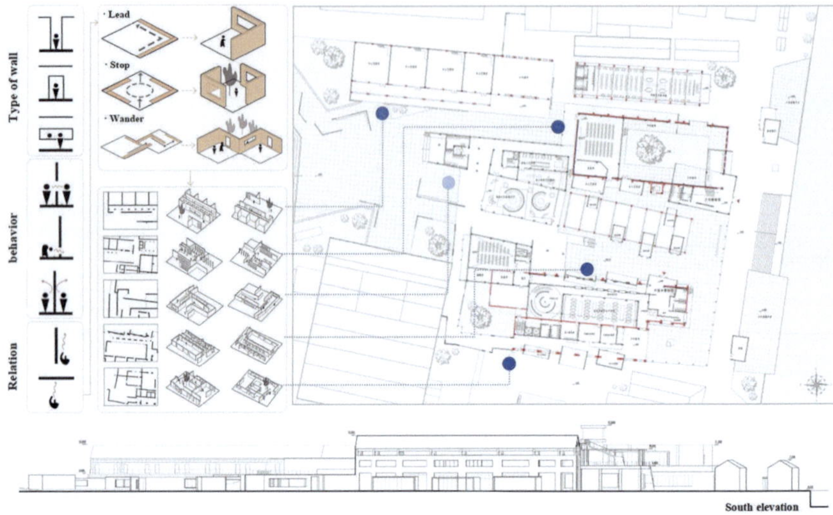

Fig. 14.13 Analysis of wall cavity types and illustration of south facade after renovation

5.2 Effectiveness Evaluation

Based on the above transformation strategies, the design intervention for Qiaoxi was implemented, resulting in a comprehensive plan for the museum cluster and surrounding urban area. It was then evaluated using the four key analytical dimensions mentioned earlier, with results normalized for a multi-dimensional comparison of spatiotemporal vitality before and after the transformation.

Effectiveness Evaluation of Functional Diversity

Guided by the concept of "city museum", functional integration was carried out, creating a highly mixed-use space where production, exhibition, and sales intermingled. Functional diversity was assessed using newly generated POI data, showing a 7.85% and 8.34% increase in daytime and nighttime diversity respectively. Additionally, the disparity between day and night diversity decrease to 97.41% of its original value (Fig. 14.14).

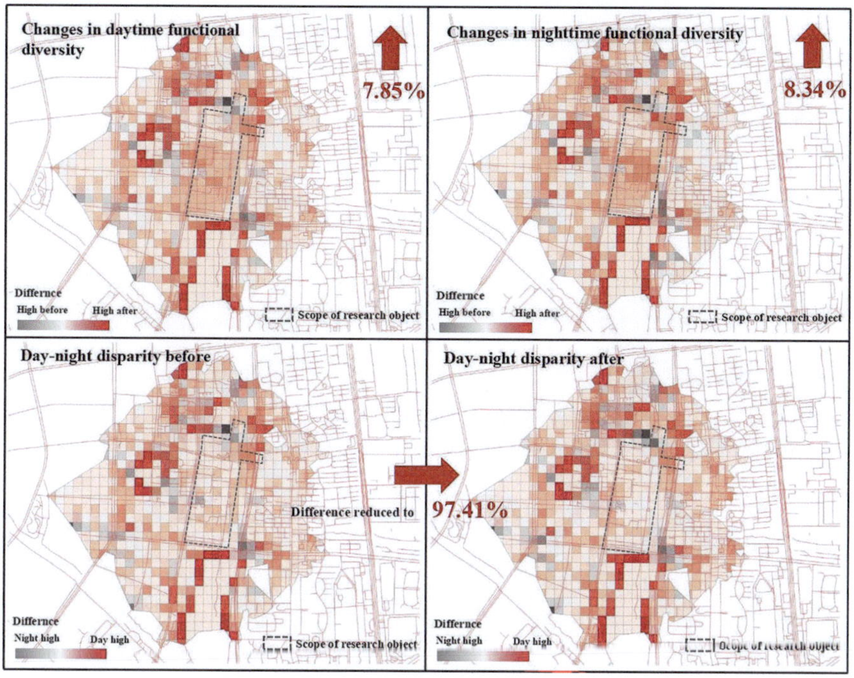

Fig. 14.14 Diagram of the changes of functional diversity before and after renovation

Effectiveness Evaluation of Supply–Demand Coupling

Guided by the concept of "space reversal", the strategy was implemented across walls, streets, and buildings, achieving nighttime visibility for most exhibition halls. Additionally, some exhibitions expanded beyond the halls through handicraft workshops and commercial functions, significantly enhancing supply–demand coupling. As shown in Fig. 14.15, the weighted average supply–demand coupling degree of six facilities types increase by 4.03%, with educational and cultural facilities rising by 6.95%. In particular, the supply–demand coupling degree of museums increase by 16.44%.

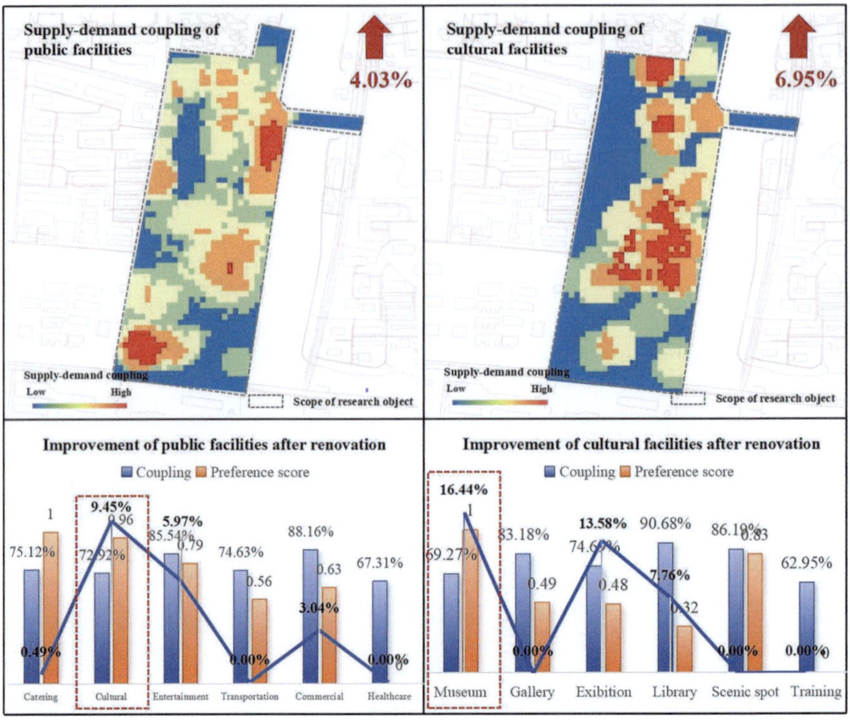

Fig. 14.15 Diagram of the changes of supply–demand coupling before and after renovation

Effectiveness Evaluation of Accessibility

Guided by the concept of "double-decker street" strategy, the museum and surrounding space interfaces were transformed. Additionally, the reconstructed street system continued the historical texture of Qiaoxi while improving road accessibility. As shown in Fig. 14.16, accessibility during the day and night has increased by 39.22% and 103.04% respectively, with the day-night disparity reduce to 82.59% of its original value.

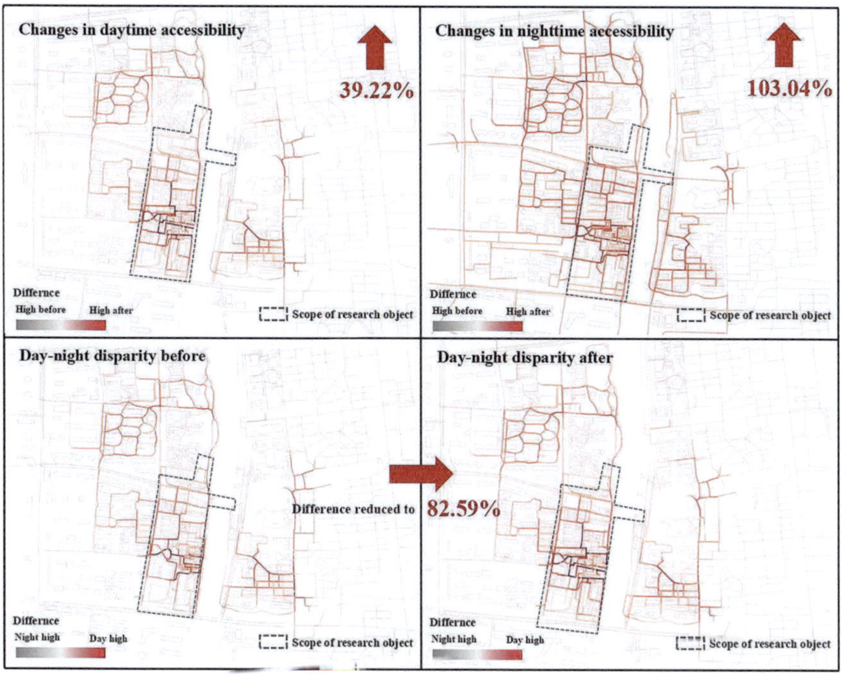

Fig. 14.16 Diagram of the changes of accessibility before and after renovation

Effectiveness Evaluation of Spatial Permeability

Guided by the concept of "facade renovation" strategy, facade and wall design changed spatial experience. Specifically, users could take advantage of the good permeability of the space to see various scenes such as exhibitions, commerce, and natural landscape, which greatly enriched the visual experience. As shown in Fig. 14.17, the overall spatial permeability of the Qiaoxi during day and night has increased by 1.46% and 3.57% respectively. Considering this indicator only obviously manifested in the renovation site or a small surrounding area, further measurements were carried out in the core area of the renovation, showing the promotion of 6.18% and 11.90% respectively, especially the day-night disparity was only 28.31% of the original.

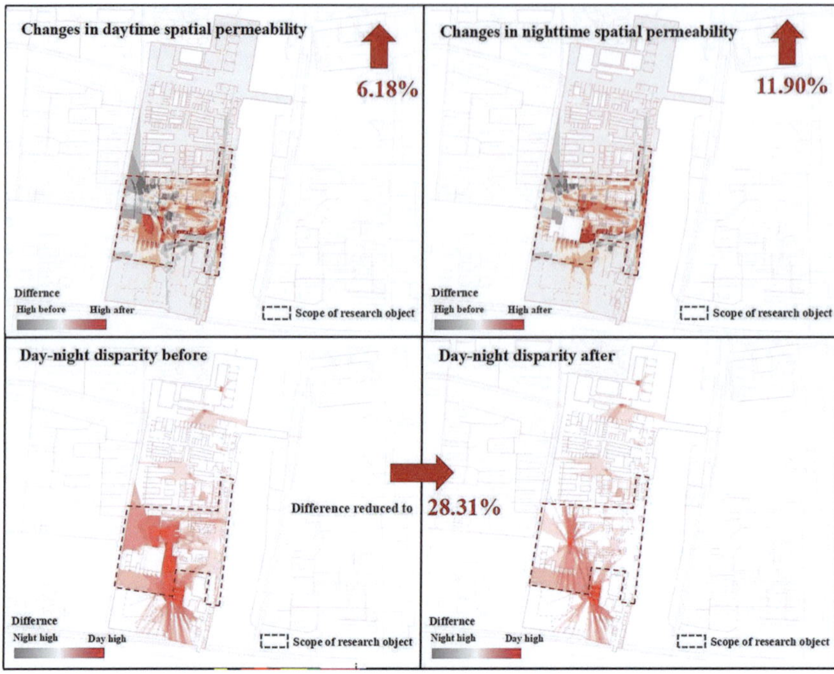

Fig. 14.17 Diagram of the changes of spatial permeability before and after renovation

Effectiveness Evaluation of Comprehensive Spatiotemporal Vitality

Based on the weight of the four key analysis dimensions affecting spatiotemporal vitality derived from the AHP in Sect. 2.2, the comprehensive measurement results and the day-night disparity of the Qiaoxi were finally calculated. As shown in Fig. 14.18, the spatiotemporal vitality during the day and night has increased by 11.40% and 21.69% respectively, with no obvious vitality gap in the central area of the district. Additionally, the vitality difference between day and night has reduced to 61.69%, while the consistency of it has increased to 1.62 times, demonstrating the effectiveness of the strategies and design interventions.

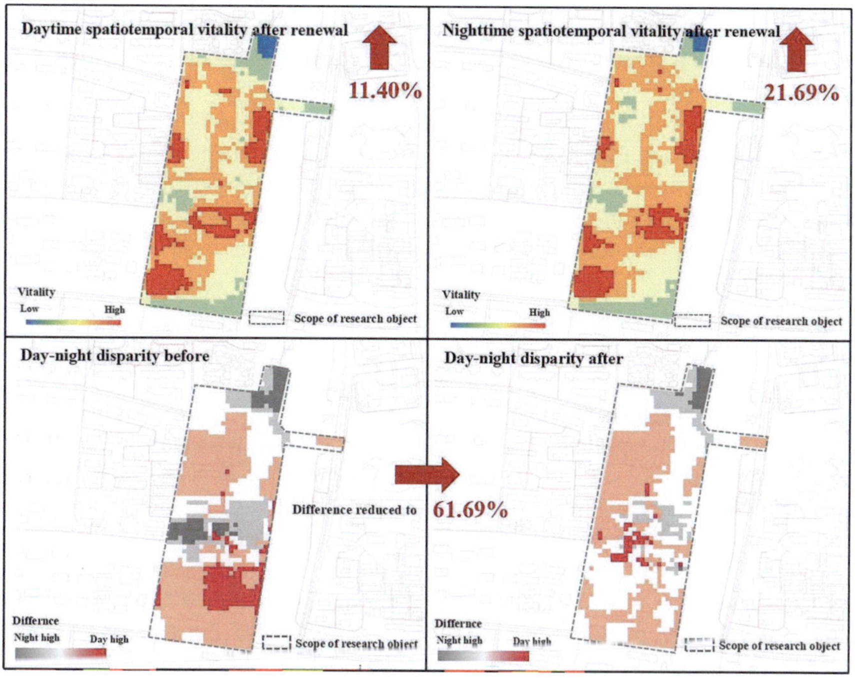

Fig. 14.18 Diagram of the changes of spatiotemporal vitality before and after renovation

6 Discussion and Conclusion

6.1 Method Innovation in Analysis and Design of Spatiotemporal Vitality of Historical and Cultural District

This study developed a digital analysis and design method for historical and cultural districts from the perspective of spatiotemporal vitality.

First, for the analytical perspective, by sorting out classic theories, the key dimensions of vitality of historical district were clarified. Meanwhile, the relationship between spatiotemporal vitality and functional characteristics, as well as spatial perception was explored, including functional diversity, supply–demand coupling, accessibility and spatial permeability. Functional diversity and supply–demand coupling had greater impact on vitality, especially in catering services and educational & cultural facilities. In contrast, accessibility and spatial permeability had a relatively small impact, but could be greatly enhanced through design interventions, thereby contributing to overall vitality improvements.

Additionally, for the research methods, using the multi-source data and new technologies, a quantitative measurement of human-centered perception in historical and cultural district was achieved. This enabled precise problem identification and analysis. Based on the weighted importance of each dimension, a comprehensive spatiotemporal vitality map was generated, highlighting key areas for improvement.

Third, for application scope, the proposed key dimensions and quantitative methods were transferable to other historical districts, offering a generalizable framework.

Specifically, this approach used digital technology to evaluate the spatiotemporal vitality of historical district, pinpointed differences and their causes, and guided targeted design interventions.

Finally, for the workflow, beyond the early stages of urban design, this study embedded quantitative research throughout the entire process, from diagnosing issues to evaluating design outcomes. A quantitative effectiveness evaluation method for the continuation and enrichment improvement of the spatiotemporal vitality had been proposed and carried out into practice, ensuring consistent development over time and space in historical districts.

6.2 Limitations

There are also several limitations to the study. First, this study mainly focus on ground-level accessibility and spatial permeability, but the impact of steps, corridors and upper-floor windows haven't been considered. In the future, three-dimensional spatial data should be included, and analitical tools like sDNA and experiments such as gaze tracking should be applied. Additionally, the key analytical dimensions need to be further completion. For instance, although physical factors are rarely emphasized in classic theories of spatial vitality, elements like sound, light, and heat at different times significantly influence spatial perception and thereby impact vitality. Endeavours should be made in future works to quantify the affect of these effects on spatiotemporal vitality in historic districts using sensors and computer vision technologies. Furthermore, some data used in this study may contain random errors. For instance, the social survey employed to calculate supply–demand coupling is limited by the scope of questionnaire distribution. In the future, integrating multiple data sources, such as mobile phone signaling, will help correct these discrepancies.

6.3 Future Applicability

This study incorporated multiple time dimensions to evaluate the vitality of historic districts while proposed a digital analysis framework and measurement method with broad applicability. Building on this, this study attempted to construct a "research-practice-feedback" application loop, offering innovative pathways for the transformation of urban design from experience-led to design science. It also provided new possibilities for precise urban design and the development of design science from the perspective of spatiotemporal dynamics.

References

1. Wang J (2021) Exploring the approaches of adaptive conservation, reconstruction and revitalization of historic districts a case study of Gunanjie Street in Dingshu. Yixing. Archit J 05:1–7. https://doi.org/10.19819/j.cnki.ISSN0529-1399.202105001
2. Sun Z et al (2023) Design of enhancing the vitality of residential historical and cultural districts based on RESIDENTS' SATISFACTION—a case of ZHONGSHAN middle road historical and cultural neighborhood in Hangzhou. Design J 8(3):1989–1999. https://doi.org/10.12677/Design.2023.83239

3. Ministry of Housing and Urban-Rural Development, National Development and ReformCommission (2024) Notice on issuing the construction guidelines for protection and enhancement projects of historical and cultural cities and blocks (Trial). GOV. https://www.gov.cn/zhengce/zhengceku/202402/content_6933895.htm
4. Ren L, Yuan D (2013) Analysis of current urban vitality and function. Architect Design 7(1):3
5. Fan J et al (2019) Multi-dimensional evaluation and guidance for quality pedestrian street space: an analysis of multi-sourced urban data. Planners 35(14):5–11. CNKI:SUN:GHSI.0.2019–14–001
6. Ye Y et al (2024) Precise urban regeneration via digital urban design: using Lujiazui water loop project as the case. The Architect 2024(02):31–38. CNKI:SUN:JZSS.0.2024-02-005
7. Qian S et al (2010) Relationship between place attachment of historic street and customer shopping satisfaction: a case study of Guanqian street. Progress Geogr 29(03):355–362. CNKI:SUN:DLKJ.0.2010–03–016
8. Hong Y (2017) Study on assessment system for modern adaptation of conservation areas of the grand Canal (Hangzhou Main Section) [Doctoral dissertation, Zhejiang University]. Doctoral Dissertations. https://kns.cnki.net/KCMS/detail/detail.aspx?dbname=CDFDLAST2018&filename=1017270687.nh
9. Zhong W, Wang D (2016) A study on the dynamic characteristics of spatial activities based on mobile phone signaling data: A case study of Zhangjiang Hi-Tech Park in Shanghai. In Proceedings of the 60th anniversary of urban planning: achievements and challenges—2016 China urban planning annual conference (Vol 04, New Technologies in Urban Planning Applications). ConferenceArticle/5af26b53c095d716587dd8ba
10. Chen Y et al (2023) Urban renewal evaluation method based on fuzzy analytic hierarchy process: a case of Qiaoxi historic block in Hangzhou section of the Canal. Urbanism Archit 20(14):82–84+99. https://doi.org/10.19892/j.cnki.csjz.2023.14.20
11. Wang D, Hu Y (2022) Urban spatial-temporal activity planning: concept. Framework and Prospect. Urban Planning Forum 2022(01):44–50. https://doi.org/10.16361/j.upf.202201006
12. Sheng Q et al (2014) Spatial conditions for targeted and optional consumption: a space syntax study on Wangfujing area and three shopping malls. Archit J 2014(06):98–103. CNKI:SUN:JZXB.0.2014-06-021
13. Ji X et al (2020) Study on the cognitive approach of historic urban landscapes from the perspective of multi-dimensional interaction. Urban Development Studies 27(07):67–74+2+33. CNKI:SUN:CSFY.0.2020–07–011
14. Banerjee T, Southworth M (1996) City sense and city design. writings and projects of Kevin Lynch [J]. Landscape J 15(2):167–168
15. Gehl J (2002) Life between buildings: Using public space (R. He, Trans.). China Architecture & Building Press
16. Jacobs J (2006) The death and life of great American cities (J. Jin, Trans.). Yilin Press
17. Lynch K (1964) The image of the city [M]. MIT Press
18. Naito A (2017) The aesthetics of townscapes. Jiangsu Phoenix Literature & Art Publishing House
19. Whyte WH (1980) The social life of small urbanspaces [J]
20. De Nadai M, Staiano J, Larcher R et al (2016) The death and life of great Italian cities: a mobile phone data perspective[C]. Proceedings of the 25th international conference on world wide web. International world wide web conferences steering committee, pp 413–423
21. Long Y, Ye Y (2016) Human-scale urban form: measurements, performances, and urban planning & design interventions. South Architecture 2016(5):41–47. CNKI:SUN:NFJZ.0.2016-05-008
22. Long Y, Zhou Y (2016) Quantitative evaluation on street vibrancy and its impact factors: a case study of Chengdu. New Architecture 2016(01):52–57. CNKI:SUN:XJZJ.0.2016-01-010

23. Jiang X et al (2022) Research on parking problems in Baoding historic blocks based on supply and demand. Shanxi Architect 48(08):34–37. https://doi.org/10.13719/j.cnki.1009-6825.2022.08.009
24. Zhan H (2016) Study on road traffic improvement of the Quanzhou ancient City based on the supply and demand [Master's Thesis, Tsinghua University]. Master's Theses. https://kns.cnki.net/KCMS/detail/detail.aspx?dbname=CMFD201801&filename=1017818034.nh
25. Wang M, Wang Y (2021) Measurement and optimization of supply and demand of park green space at urban block scale: a case study of Yangpu District, Shanghai. Landscape Architecture, 28(2):22–27. https://doi.org/10.14085/j.fjyl.2021.02.0022.06
26. Duan X (2021). Research on the measurement and promotion strategy of New City Vitality based on big data—a case study in wuhan metropolitan area [Master's Thesis, Huazhong University of Science and Technology]. Master's Theses. https://doi.org/10.27157/d.cnki.ghzku.2021.003490
27. Cui Y et al (2016) Based on space syntax to explore the rehabilitation of historic district under the influence of chongqing early industry—a case study Of Shiba Ti central bazaar and 2# street. Architect Culture 2016(04):112–114. CNKI:SUN:JZYW.0.2016-04-032
28. Sarkar C, Webster C, Pryor M et al (2015) Exploring associations between urban green. street design and walking: results from theGreater London boroughs [J]. Landscape and UrbanPlanning 143:112–125
29. Sevtsuk A, Mekonnen M (2012) Urban network analysis [J]. Revue internationale de géomatique-n 287:305
30. Zeng C, Song Y, He Q et al, Spatially explicit assessment on urban vitality: case studies
31. Dadashpoor H, Rostami F (2017) Measuring spatial proportionality between service availability, accessibility and mobility: Empirical evidence using spatial equity approach in Iran [J]. J Transp Geogr 65:44–55
32. Ye Y et al (2019) Human-scale quality on streets: a large-scale and efficient analytical approach based on street view images and new urban analytical tools. Urban Planning Int 34(1):18–27. CNKI:SUN:GWCG.0.2019–01–003
33. Wu P et al (2023) Study on matching efficiency of supply and demand of public service function facilities in Shanghai: analysis based on the heterogeneity characteristics of residents. Urban Probl 2023(04):87–95. https://doi.org/10.13239/j.bjsshkxy.cswt.230409

Computational Landscape

Thermal Comfort Optimization in Renovated Courtyards of Primary and Secondary Schools in Lingnan Area Based on Performance Simulation: A Case Study in Shenzhen

Yuqian Mao, Gang Yu, Xingjie Zhu, and Xiaoyu Lin(✉)

School of Architecture, Harbin Institute of Technology (Shenzhen), Shenzhen, China
linxiaoyu@hit.edu.cn

Abstract. With the advancement of China's economic development and policy incentives, educational standards for primary and secondary schools have significantly improved in recent years. Many campuses built in Shenzhen during the early 2000s can no longer meet the growing demands of education. Against urban renewal and high-quality development, renovating ageing school buildings in Shenzhen is urgent. As a typical architectural form in Lingnan school campuses, courtyards serve various outdoor teaching, communication, and extracurricular activities due to their advantageous location and rich environmental configuration. They are crucial for improving campus quality in subtropical areas. Different renovation strategies for school courtyards have varying impacts on outdoor thermal comfort. This study selected representative ageing school campuses in Shenzhen for typological analysis and extracted a prototype of Lingnan school courtyards as the research object. ENVI-met was used to simulate the outdoor thermal environment of typical courtyards, combining qualitative and quantitative approaches to analyze the impacts of four passive renovation strategies—opening, shading, inner facades and underlying surfaces—on outdoor thermal comfort. A comparative analysis was conducted to determine optimal renovation strategies, providing a reference for courtyard renovation and thermal environment optimization on Lingnan school campuses.

Keywords: Lingnan area · Primary and secondary schools · Courtyard renovation · Performance simulation · Thermal comfort

Nomenclature

d	Length of openings
L	Length of the courtyard's one side
ρ	Shading grid density
S	Courtyard area
s_1	Horizontal shading area
s_2	Grass-covered area
p	p-Value

J. Fu et al. (eds.), *AI for Architecture*, Computational Design and Robotic Fabrication,
https://doi.org/10.1007/978-981-95-0974-4_15

PET Physiologically Equivalent Temperature (°C)
mPET Mean Physiological Equivalent Temperature (°C)
T_a Air Temperature (°C)
RH Relative Humidity (%)
WS Wind Speed (m/s)
T_{mrt} Mean Radiation Temperature (%)

1 Introduction

In the Lingnan region of China, summers are long, hot, and humid. Courtyards, as a climate-responsive design, are widely applied in the layout of primary and secondary school campuses. As outdoor public spaces link teaching and extracurricular activities, the thermal comfort of courtyards significantly affects the willingness of teachers and students to stay and engage in activities and communication in these areas. Thermal comfort refers to the state of being satisfied with the thermal environment [1], influenced by physical factors (temperature, relative humidity, wind speed, and mean radiant temperature) and personal factors (metabolic rate and clothing insulation) [2] according to the existing researches.

With economic growth and elevated teaching standards, numerous ageing courtyard-style school buildings in Lingnan area urgently need renovation. Courtyard optimization can improve the thermal environment effectively, primarily influenced by physical factors such as air temperature, relative humidity, solar radiation, and wind speed [3]. Architectural renovation measures mainly impact courtyard's thermal environment and comfort by altering solar radiation and wind flow [4]. Therefore, improving campus quality through passive renovation strategies for courtyard thermal environment is effective. Courtyard thermal environment is significantly affected by features such as geometric shape, orientation, wall materials, underlying surface materials and landscaping elements [5]. The geometry and orientation of the courtyard influence lighting and wind environments [6]. High-reflectance facades and paving materials can reduce solar radiation absorption, cooling the courtyard [7]. Landscaping elements such as vegetation improve the thermal environment through shading, wind shielding, and evapotranspiration [8]. For courtyard-style campus renovations, geometric shapes and orientations are relatively fixed. Given economic costs and construction duration constraints, practical and standard measures include creating opening, adding shading components, changing facade materials, altering courtyard pavements, and reorganizing vegetation.

Current research on school courtyard thermal environments primarily focuses on geometric control in the design stage and comparisons of thermal comfort before and after renovation. However, there is a lack of in-depth discussion on the relationship between courtyard renovation strategies and outdoor thermal comfort. Therefore, this study established a typical courtyard model for Lingnan schools, employing environmental simulation and controlled variables to qualitatively and quantitatively analyze the effects of four renovation measures (opening, shading, inner facades and underlying surfaces). It explores the correlation between renovation strategies, environmental physical factors and thermal comfort, proposing optimization strategies for renovating and improving the thermal environment of Lingnan school courtyards.

2 Methodology

Figure 15.1 illustrates the research framework for optimizing the thermal comfort of courtyard renovations in primary and secondary schools in Lingnan area based on performance simulation technology. This study establishes a prototype of Lingnan school courtyards and proposes four passive renovation strategies. Using performance simulation software (ENVI-met) and controlled variable methods, the outdoor thermal environment of the prototype and its renovation models were simulated and evaluated during specific periods. The PET distribution and mean PET (mPET) of the courtyard observation area were selected as comprehensive indicators for evaluating thermal comfort. A combination of qualitative and quantitative analysis was employed to assess the effects and impacts of different renovation strategies. Additionally, statistical methods were used to compare the relationships between renovation strategies, environmental physical factors, and thermal comfort, ultimately deriving optimized renovation strategies for improving thermal comfort in Lingnan school courtyards.

2.1 Case Study

Courtyard Prototype

Under the influence of teaching demands and regulatory constraints, courtyard-style buildings in Lingnan primary and secondary schools have common construction patterns. Using Shenzhen as a representative city in the Lingnan region, this study selected 30 primary and secondary school samples built around 2000 in different parts of the city. Statistical analysis of these school courtyard forms, dimensions, orientations, and story heights was conducted to summarize their general characteristics. As shown in

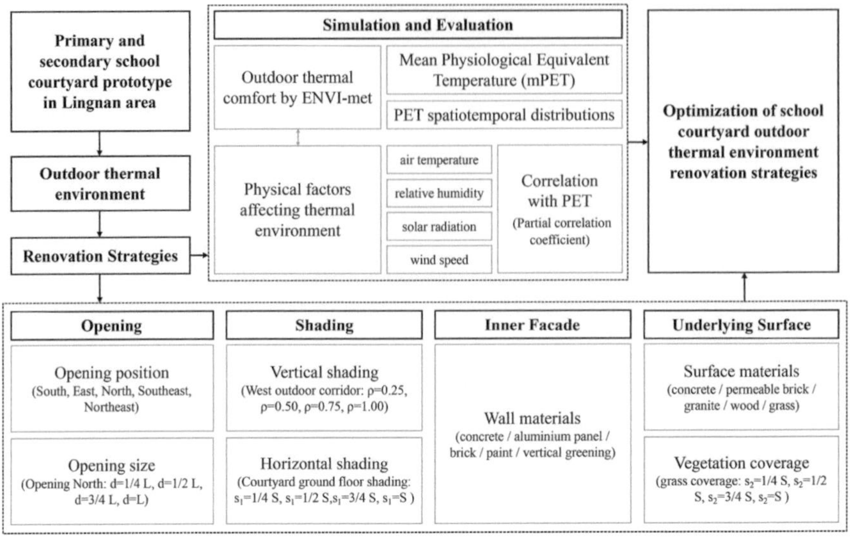

Fig. 15.1 Research framework

Table 15.1, Lingnan school courtyards can be categorized into three forms: "Closed Square" "U-shaped" and "U-shaped with Corridors" with the latter being the most widely used. The average courtyard length is approximately 58 m, and the width is 45 m, as the length-to-width ratio is about 0.8. Most school buildings are 4–5 stories high and predominantly oriented north–south. Based on these results, combined with structural design and relevant standards (e.g., "Design Code for Primary and Secondary Schools"), the basic courtyard prototype was defined as a north–south oriented "U-shaped with Corridors" form, with a building height of 20 m, length of 60 m, width of 48 m, and courtyard dimensions of 48 m × 28 m. This area was defined as the observation zone for subsequent thermal environment simulations (Fig. 15.2). The walls of the prototype were modelled as concrete, and the courtyard paving material was set as cement.

Table 15.1 Summary of courtyard forms in typical primary and secondary school samples in Shenzhen

Courtyard forms	Typical primary and secondary school samples in Shenzhen
Closed square	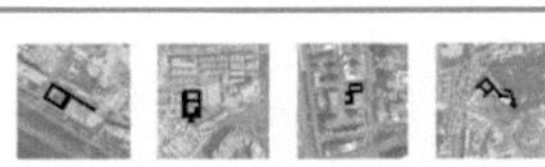
U-shaped	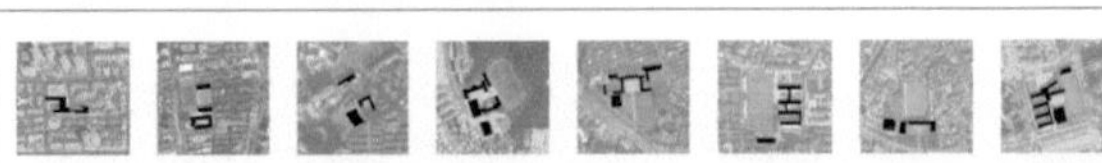 

(*continued*)

Table 15.1 (*continued*)

Courtyard forms	Typical primary and secondary school samples in Shenzhen
U-shaped with corridors	

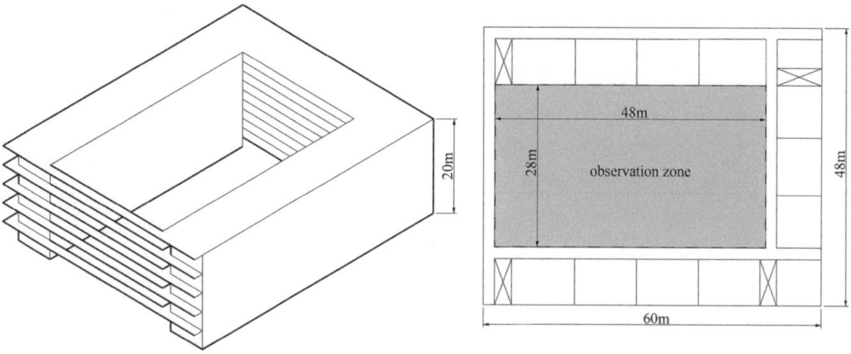

Fig. 15.2 Diagram of the typical courtyard prototype of primary and secondary schools in Lingnan area (left) and the thermal environment simulation observation zone (right)

Renovation Strategies

The renovation of ageing school buildings focuses on minimal interventions to existing structures to minimize economic costs and shorten construction periods. Considering the courtyard's geometric shape, facade materials, ground paving, and landscaping elements, four primary renovation targets were identified: opening, shading, inner facades, and underlying surfaces.

Based on these targets, renovation measures were further subdivided by design elements: opening contained position and size of openings; shading was divided into vertical shading and horizontal shading; inner facades referred to facade materials; underlying surfaces included underlying surface materials and vegetation coverage. Table 15.2 summarizes the courtyard prototype variations corresponding to these renovation strategies.

Table 15.2 Four types of passive renovation strategies for courtyards

Renovation strategy	Prototype diagram	Design element	Courtyard prototype variation diagram
Opening	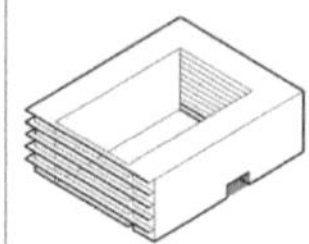	Opening position (A)	 A-1 South, A-2 East, A-3 North, A-4 Southeast, A-5 Northeast
		Opening size (B)	 B-1 d=1/4L, B-2 d=1/2L, B-3 d=3/4L, B-4 d=L
Shading	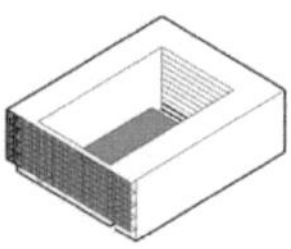	Vertical shading (C)	 C-1 ρ=0.25, C-2 ρ=0.50, C-3 ρ=0.75, C-4 ρ=1.00
		Horizontal shading (D)	 D-1 s_1=1/4S, D-2 s_1=1/2S, D-3 s_1=3/4S, D-4 s_1=S
Inner facade	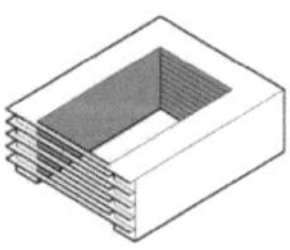	Wall materials (E)	 E-1 Aluminium panel, E-2 Brick, E-3 Paint, E-4 Vertical greening
Underlying surface	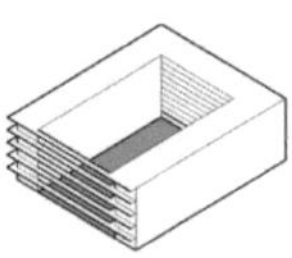	Surface materials (F)	 F-1 Permeable brick, F-2 Granite, F-3 Wood, F-4 Grass
		Vegetation coverage (G)	 G-1 s_2=1/4S, G-2 s_2=1/2S, G-3 s_2=3/4S, G-4 s_2=S

2.2 Thermal Environment Simulation

Simulation Setup

The summer climate is hot and humid in Lingnan area, lasting up to eight months. With global warming intensifying, extreme heat events have become more frequent. Therefore, improving outdoor thermal environments in Lingnan during summer is paramount. Considering the typical school schedules, the simulation period was set for June, focusing on the summer solstice (June 21) between 8:00 and 18:00.

ENVI-met was used for thermal environment simulation, comprehensively considering meteorological factors and their interactions. It has been validated by calibrating measured and simulated values, making it suitable for small-scale courtyard microclimate simulations. Table 15.3 outlines the simulation conditions for the courtyard prototype and renovation schemes (specific meteorological conditions referenced in local EPW files). As primary and secondary school courtyards mainly serve student activities, the simulation object was set as children (10-year-old males wearing summer clothing), and the thermal comfort index PET was calculated at the height of 1.4 m.

Table 15.3 ENVI-met simulation conditions

Shenzhen	22.55N;114.10E (UTC + 8)
Simulation time	8:00 to 18:00, June 21, 2023
Model dimensions and grid settings	X-80 Y-75 Z-20 (dx = 2 m, dy = 2 m, dz = 1 m)
Wind speed and direction (N:0°, S:180°)	1–4 m/s, 20°–160°
Air temperature	28.5–31.4°C
Humidity	19.9–24.8 g/Kg

Performance Indicators

The thermal environment of courtyards directly affects the thermal comfort of outdoor users. Physiological Equivalent Temperature (PET) is a commonly used indicator for evaluating outdoor thermal environments. Influenced by factors such as climate, environment and thermal adaptation, PET requires adjustments based on specific climatic conditions to ensure accurate evaluation. Table 15.4 presents the thermal sensation ranges for PET in Lingnan area during summer based on existing studies [9].

Table 15.4 Thermal sensation ranges for PET in Lingnan area during summer

Sensation	Cold	Cool	Slightly cool	Neutral	Slightly warm	Warm	Hot
PET (°C)	<11.3	11.3–19.2	19.2–24.6	24.6–29.1	29.1–36.3	36.3–53.6	>53.6

However, due to the extremely hot and humid outdoor conditions in Lingnan during summer, PET often falls outside the comfortable range, clustering predominantly in the "hot" category. Therefore, this study chose the mean PET (mPET) of the courtyard observation area as the evaluation metric for thermal environment optimization. Statistical characteristics and spatiotemporal distributions of PET were analyzed to assess thermal comfort improvements.

3 Results and Discussion

Figure 15.3 shows the average thermal comfort variation of the courtyard prototype on the summer solstice. Throughout the day, the mean PET (mPET) in the courtyard remained high, generally exhibiting a rising trend followed by a decline. Considering the typical schedule of outdoor activities in primary and secondary schools, 3:00 PM—when the courtyard PET reaches its peak—was selected for a detailed comparative analysis of how different renovation strategies impact courtyard thermal comfort.

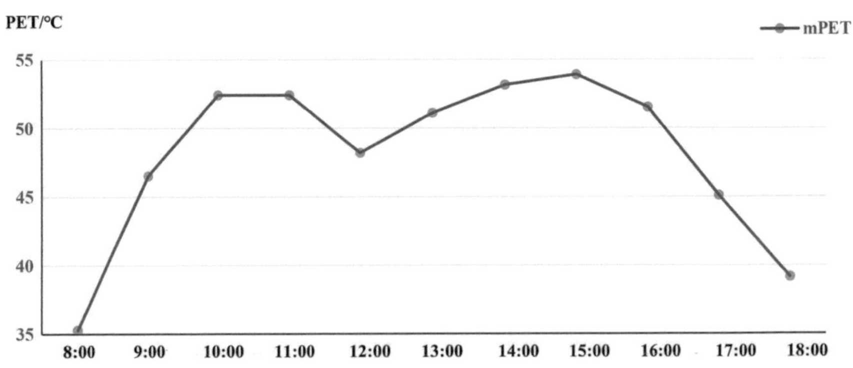

Fig. 15.3 Average thermal comfort variation characteristics of the courtyard prototype on the summer solstice

Physical factors such as air temperature, relative humidity, solar radiation, and wind speed influence the PET in the courtyard while interacting with one another. Therefore, partial correlation analyses were conducted between the PET results of different renovation strategies and these physical factors to determine how each strategy affects courtyard thermal comfort. Since air temperature (T_a) and relative humidity (RH) are strongly coupled (correlation coefficient approximately -1), only relative humidity was included in the analysis.

3.1 Opening

The results of the thermal comfort simulation for the courtyard prototype and two types of opening strategies (A and B) at 15:00 on the summer solstice are shown in Fig. 15.4. It is visually apparent that the PET in courtyard areas near the openings significantly decreased, improving thermal comfort. Figure 15.5 illustrates the PET distribution in the courtyard observation area for different opening positions (A) and opening sizes (B). All modified scenarios showed a reduction in mPET in group A, with the south-side opening (A-1) having the lowest mPET but a more dispersed PET distribution, indicating uneven thermal comfort. This may be due to the dominant southeast wind direction intensifying internal airflow instability with the south-side opening. In contrast, the east-side opening (A-2) not only had a lower mPET but also exhibited a more uniform PET distribution, resulting in more consistent thermal comfort improvements. Among the three east-side openings (A-2, A-4, A-5), the central opening (A-2) yielded the lowest mPET and a more uniform PET distribution, suggesting that a centrally located opening has a more direct impact on air flow and thus improves thermal comfort more effectively than corner openings.

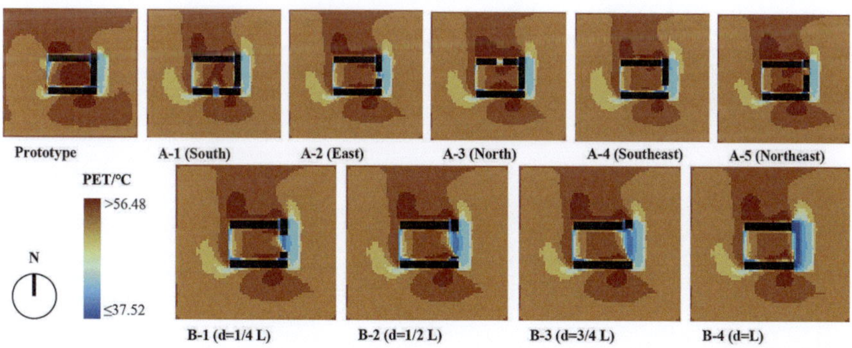

Fig. 15.4 Simulation results of thermal comfort in courtyard with opening renovation strategies

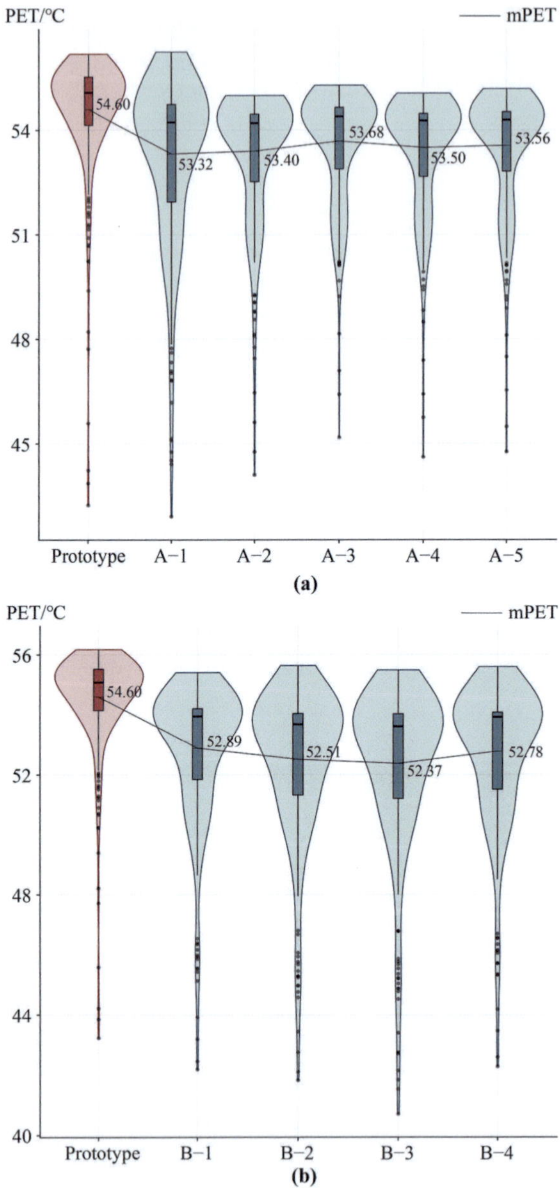

Fig. 15.5 PET violin diagram of opening renovation strategies (A and B)

In group B, the mPET reduction is more significant, and the PET distribution is more dispersed than in group A. This indicates that larger openings increased the effective cooling area, leading to uneven thermal comfort. As the opening size increased, mPET decreased, reaching its minimum at 3/4 of the courtyard's east-side length (B-3) before slightly increasing with further extension. This suggests that a moderate opening length optimally balances thermal comfort improvements and space efficiency.

Partial correlation analysis (Table 15.5) shows that wind speed and solar radiation significantly influenced PET ($p < 0.01$). Wind speed is negatively correlated with PET—higher wind speed lowered PET, enhancing thermal comfort—while solar radiation positively correlates with PET. Solar radiation had a more significant overall impact on PET than wind speed. However, wind speed exhibited greater variability, suggesting that changes in opening position influence PET primarily by affecting airflow. Relative humidity has an insignificant effect on PET, likely due to consistently high summer humidity in Lingnan area. Therefore, wind speed and solar radiation are identified as the dominant factors affecting PET.

Table 15.5 Partial correlation coefficient between PET and physical factors in courtyard area of opening renovation strategies (Note: * means$p < 0.01$)

Analysis variable	Control variables	Prototype	A-1	A-2	A-3	A-4	A-5
RH	WS&T_{mrt}	−0.167	−0.202	−0.249	−0.327	−0.001	−0.165
WS	RH&T_{mrt}	−0.774*	−0.786*	−0.612*	−0.391*	−0.808*	−0.789*
T_{mrt}	RH&WS	0.892*	0.888*	0.606*	0.463*	0.892*	0.885*

Analysis variable	Control variables	Prototype	B-1	B-2	B-3	B-4
RH	WS&T_{mrt}	−0.167	0.219	0.228	0.092	0.127
WS	RH&T_{mrt}	−0.774*	−0.861*	−0.862*	−0.842*	−0.816*
T_{mrt}	RH&WS	0.892*	0.889*	0.903*	0.897*	0.905*

3.2 Shading

Figure 15.6 displays the simulation results for the two shading strategies (C and D) during typical high-temperature periods in summer. PET values decreased across the courtyard, with a significant drop in shaded areas. Comparing the mPET values in Fig. 15.7, horizontal shading (D) consistently outperformed vertical shading (C) in reducing PET. In group C, increasing the density of vertical shading grids on the west side had little impact on mPET but stabilized PET distribution. The limited shaded area during afternoon hours reduced the effectiveness of vertical shading. However, group D demonstrated that increasing horizontal shading coverage initially decreased mPET, reaching its lowest point when shading covered half the courtyard (D-2). Beyond this point, further shading reduced airflow, causing mPET to rise and decreasing overall thermal comfort.

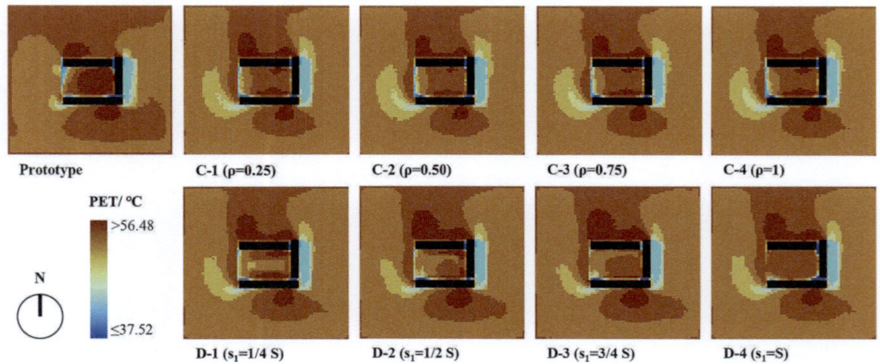

Fig. 15.6 Simulation results of thermal comfort in courtyard with shading renovation strategies

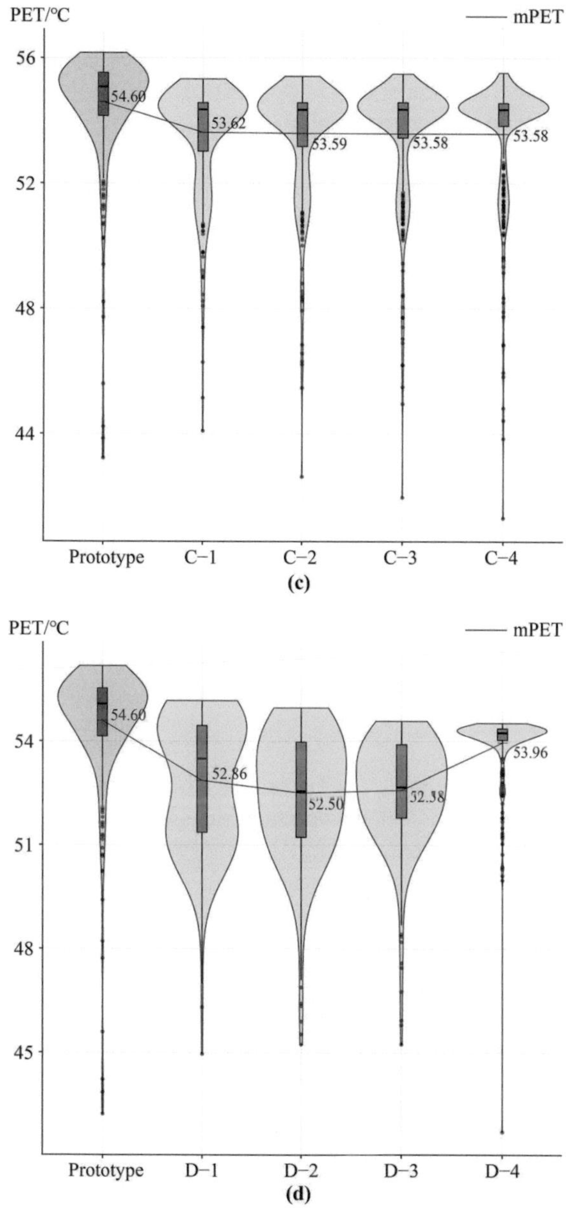

Fig. 15.7 PET violin diagram of shading renovation strategies (C and D)

Partial correlation analysis (Table 15.6) revealed that solar radiation impacted PET more than wind speed in group C. As shading density increased, the influence of solar radiation on PET decreased, confirming that vertical shading improves thermal comfort by reducing solar radiation. Meanwhile, horizontal shading significantly altered both wind flow and solar radiation in group D. As horizontal shading coverage expanded, the partial correlation coefficient of wind speed gradually increased from less than solar radiation, and wind speed became the dominant factor affecting PET. However, when shading exceeded three-quarters of the courtyard (D-3), airflow decreased, causing PET to rise and reducing overall thermal comfort.

Table 15.6 Partial correlation coefficient between PET and physical factors in courtyard area of shading renovation strategies (Note: * means$p < 0.01$)

Analysis variable	Control variables	Prototype	C-1	C-2	C-3	C-4
WS	RH&T_{mrt}	-0.774^{*}	-0.725^{*}	-0.720^{*}	-0.719^{*}	-0.735^{*}
T_{mrt}	RH&WS	0.892^{*}	0.720^{*}	0.760^{*}	0.797^{*}	0.845^{*}
Analysis variable	Control variables	Prototype	D-1	D-2	D-3	D-4
WS	RH&T_{mrt}	-0.774^{*}	-0.798^{*}	-0.839^{*}	-0.895^{*}	-0.798^{*}
T_{mrt}	RH&WS	0.892^{*}	0.909^{*}	0.763^{*}	0.726^{*}	0.635^{*}

3.3 Inner Facade

The simulation results for different inner facade materials (E) are shown in Fig. 15.8. PET values decreased near the modified facade surfaces. The ranking of mPET from lowest to highest was: vertical greening < brick < paint < aluminium panel < concrete. Vertical greening (E-4) was the most effective, demonstrating that materials with high reflectivity and thermal mass reduce solar heat absorption, lower surface temperature, and minimize heat transfer between the building and courtyard. Additionally, vegetation enhances cooling through evapotranspiration. However, the PET distribution became more uneven after renovation, suggesting that facade materials have a limited impact on distant courtyard areas.

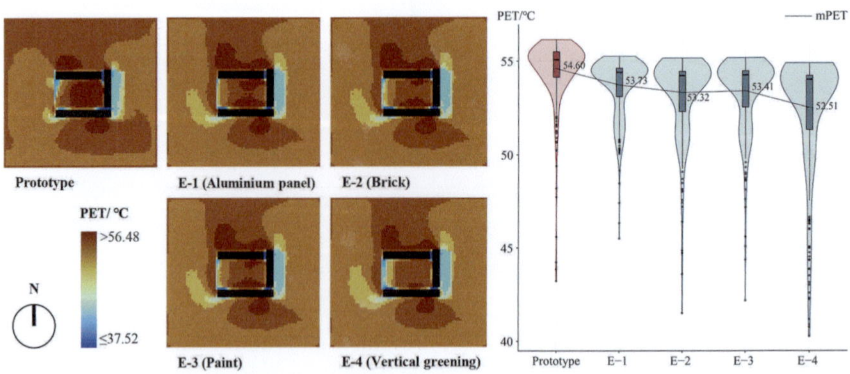

Fig. 15.8 Simulation results and PET violin diagram of thermal comfort in courtyard with inner facades renovation strategies (E)

In addition, Table 15.7 indicates that solar radiation had a more decisive influence on PET than wind speed. The partial correlation coefficient of solar radiation in group E is greater than that of wind speed, and the fluctuation is more substantial. This further confirms that inner facade materials primarily affect thermal comfort by altering solar absorption and heat transfer.

Table 15.7 Partial correlation coefficient between PET and physical factors in courtyard area of inner facades renovation strategies (Note: * means$p < 0.01$)

Analysis variable	Control variables	Prototype	E-1	E-2	E-3	E-4
WS	RH&T_{mrt}	−0.758*	−0.733*	−0.868*	−0.739*	−0.733*
T_{mrt}	RH&WS	0.860*	0.828*	0.907*	0.849*	0.828*

3.4 Underlying Surface

Figure 15.9 shows the simulation results for different underlying materials (F) and vegetation coverage (G). PET values in the central courtyard dropped significantly after renovation compared to the prototype. Figure 15.10 further illustrates a downward shift in PET distribution and a notable decrease in mPET, confirming that underlying surface changes effectively improve thermal comfort.The mPET ranking from lowest to highest was: grass < granite < permeable brick < wood < concrete in group F, proving different properties of the underlying surface materials have various degrees of effects on the thermal environment of the courtyard. Grass had the best cooling effect due to not only reducing solar radiation but also evapotranspiration and soil heat exchange. Other materials mainly reduced PET by reflecting solar radiation—higher reflectivity meant less heat absorption.In group G, mPET remained stable until vegetation coverage exceeded 1/4 of the courtyard(G-1), after which mPET improvements became noticeable. However, further coverage increases showed diminishing returns.

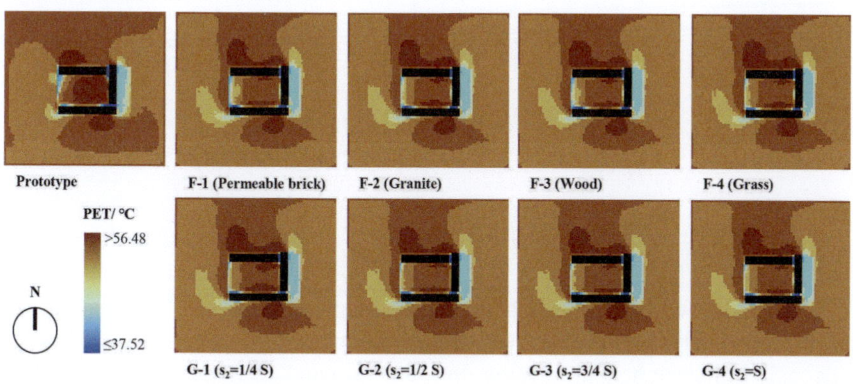

Fig. 15.9 Simulation results of thermal comfort in courtyard with underlying surface renovation strategies

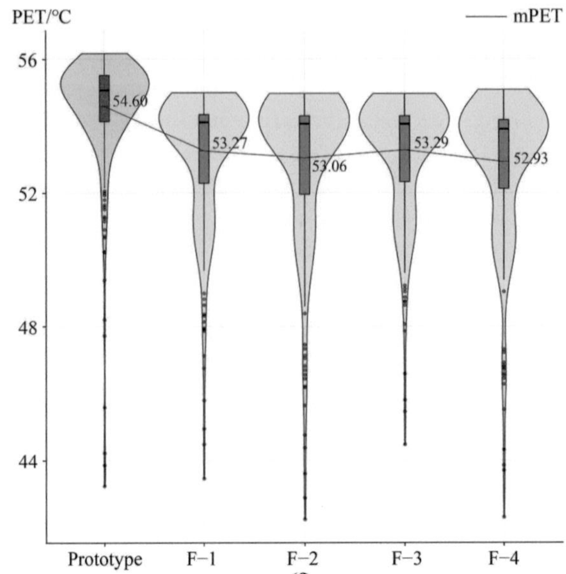

(f)

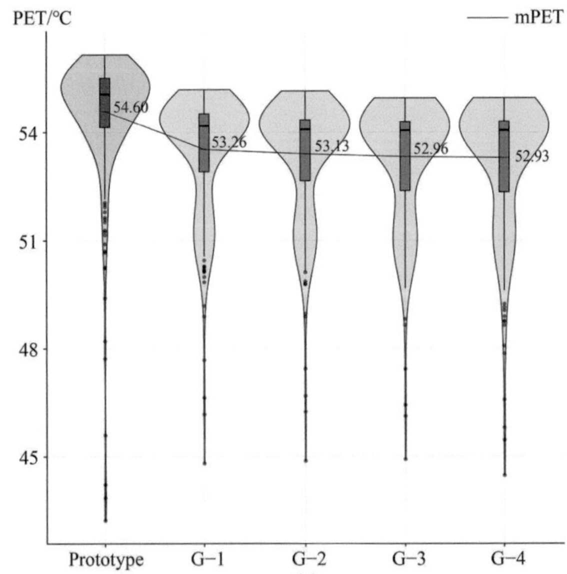

(g)

Fig. 15.10 PET violin diagram of underlying surface renovation strategies (F and G)

Table 15.8 shows that solar radiation substantially impacted PET more than wind speed in group F. This is likely because hard underlying surfaces minimally affect airflow but strongly influence heat absorption. In group G, wind speed became more influential on PET as vegetation coverage grew. This suggests that while vegetation reduces solar radiation, excessive coverage can also slow airflow, balancing out thermal comfort improvements.

Table 15.8 Partial correlation coefficient between PET and physical factors in courtyard area of underlying surface strategies (Note: * means$p < 0.01$)

Analysis variable	Control variables	Prototype	F-1	F-2	F-3	F-4
WS	RH&T_{mrt}	-0.774^*	-0.755^*	-0.875^*	-0.743^*	-0.736^*
T_{mrt}	RH&WS	0.892^*	0.804^*	0.759^*	0.855^*	0.855^*
Analysis variable	Control variables	Prototype	G-1	G-2	G-3	G-4
WS	RH&T_{mrt}	-0.774^*	-0.736^*	-0.770^*	-0.839^*	-0.808^*
T_{mrt}	RH&WS	0.892^*	0.855^*	0.866^*	0.889^*	0.882^*

4 Conclusions

This study used ENVI-met simulations to evaluate the thermal comfort performance of various courtyard renovation strategies for primary and secondary schools in Lingnan area. By analyzing PET distribution characteristics and mPET values, the study qualitatively and quantitatively assessed the thermal comfort effects of four passive design strategies. The relationship between renovation strategies, environmental physical factors, and thermal comfort was explored, resulting in the following key findings:

1. The position and size of opening, shading placement and density, inner facade materials, underlying surface materials and vegetation coverage significantly impact courtyard thermal comfort. These factors primarily influence thermal comfort by altering wind speed and solar radiation—with wind speed negatively correlated and solar radiation positively correlated with PET.
2. In summer conditions dominated by southeast winds, opening on the south or east side of the courtyard significantly improves airflow and thermal comfort. Openings that span 1/4 to 3/4 of the courtyard's one-side length achieve optimal results.
3. Horizontal shading is more effective than vertical shading in reducing PET during the afternoon. West-facing vertical shading with a 0.25 to 0.50 density and horizontal shading covering 1/4 to 1/2 of the courtyard area can significantly reduce solar radiation and enhance thermal comfort. However, exceeding this range yields diminishing returns.
4. High-reflectance facade and underlying surface materials help reduce solar radiation absorption, improving thermal comfort. Vertical greening and grass are the most effective strategies. For balanced airflow and daily usability, grass coverage of 1/4 to 1/2 of the courtyard area is recommended.

5 Limitation

This study is about summer time simulation. The following study shall focus on the remaining seasons, especially winter, to further investigate the different impact of renovation strategies on the thermal comfort of courtyards, considering year-round appropriate strategies.

Acknowledgements. This research was supported by National Natural Science Foundation Youth Project (No: 52208018), Guangdong Basic and Applied Basic Research Foundation (No: 2023A1515011351), and the Ideological and Political Course Special Project of Harbin Institute of Technology (Shenzhen) (No: HITSZIP22014).

References

1. Höppe P (1999) The physiological equivalent temperature–a universal index for the biometeorological assessment of the thermal environment. Int J Biometeorol 43(2):71–75
2. Jamei E, Rajagopalan P, Seyedmahmoudian M et al (2016) Review on the impact of urban geometry and pedestrian level greening on outdoor thermal comfort. Renew Sustain Energy Rev 54:1002–1017
3. Lai D, Liu W, Gan T et al (2019) A review of mitigating strategies to improve the thermal environment and thermal comfort in urban outdoor spaces. Sci Total Environ 661:337–353
4. Nasrollahi N, Hatami M, Khastar S et al (2017) Numerical evaluation of thermal comfort in traditional courtyards to develop new microclimate design in a hot and dry climate. Sustain Cities Soc 35:449–467
5. Zhu J, Feng J, Lu J et al (2023) A review of the influence of courtyard geometry and orientation on microclimate. Build Environ 236:110269
6. Aldawoud A (2008) Thermal performance of courtyard buildings. Energy Build 40(5):906–910
7. Chatzidimitriou A, Yannas S (2015) Microclimate development in open urban spaces: the influence of form and materials. Energy Build 108:156–174
8. Zhao Y, Chen Y, Li K (2022) A simulation study on the effects of tree height variations on the façade temperature of enclosed courtyard in North China. Build Environ 27(B):108566
9. Fang Z, Feng X, Xu X et al (2019) Investigation into outdoor thermal comfort conditions by different seasonal field surveys in China. Guangzhou. Int J Biometeorol 63(10):1357–1368

Computational Media and Art

In the Context of Metaverse: Text to Animated Axonometric Space of Ancient Cities Generation

Mengyao Li[1], Zheng Zhang[2,3], Xin Yan[4(✉)], Keyang Tang[3], and Jiaxin Ke[1]

[1] School of Software and DUT—RU International School of Information Science & Engineering at DUT, Dalian University of Technology, Dalian, China
lmy@dlut.edu.cn, kjx20000616@mail.dlut.edu.cn
[2] School of Architecture, Tsinghua University, Beijing, China
zhangzhe23@mails.tsinghua.edu.cn
[3] Future Laboratory, Tsinghua University, Beijing, China
tangkeyang@mail.tsinghua.edu.cn
[4] School of Architecture and Urban Planning, Beijing University of Civil Engineering and Architecture, Beijing, China
yanxin@bucea.edu.cn

Abstract. Given the rapid development of 5G transmission, information technology, and the evolution of people's usage habits, there has been a significant increase in demand for visual communication, primarily in the form of videos and animated graphics. Coupled with advancements in AR/VR devices, this trend is gradually transitioning towards the era of Internet 3.0 and the metaverse. The demand for mapping text into images/videos in cyberspace is consequently growing. Therefore, it is of great significance to explore a low-cost pathway for generating a textual mapping metaverse space. Drawing inspiration from the three-dimensional spatial axonometric drawing techniques in architecture, this paper addresses the lack of image-based historical records in the study of ancient cities. It proposes a method for generating vectorized, axonometric images from text, with clear scene relationships. Furthermore, by introducing the concept of "narrative", the generated scene graphs are extrapolated into animations, providing a technical pathway for lightweight generation of metaverse scenes from textual content.

Keywords: Metaverse · Text-to-image generation · Scene graph · Space form · Chinese ancient city

1 Introduction

1.1 Background

As Dr. Yanfeng Li points out in his doctoral thesis "*A Study on the Relationship between Tropes and Chinese Painting History*", there is a profound connection between Chinese painting and language, especially in literati painting, which is considered the main subject of traditional Chinese painting, reflecting the fusion of language and imagery in both content and form. In the field of text-to-image generation, a lot of research focuses on establishing an interpretable mapping from low-dimensional textual semantic information to

J. Fu et al. (eds.), *AI for Architecture*, Computational Design and Robotic Fabrication,
https://doi.org/10.1007/978-981-95-0974-4_16

high-dimensional visual information. This cross-modal task maturity represents a trend of interdisciplinary fusion between computer vision and natural language processing. Widely applied techniques include the Generative Adversarial Network (GAN) proposed [13], contrastive language-image pre-training (e.g., CLIP, represented by DALL-E) [12], diffusion model architectures (e.g., Stable diffusion) [14], among others.

However, in the context of the metaverse, current text-to-image generation technologies face challenges in generating complex scene images with multiple interacting graphic objects. Models struggle to directly parse the interaction relationships and positional information between targets from the semantic features of the text. Beyond conventional Natural Language Processing (NLP) semantic analysis, delving into the essence of narratives and studying the spatial scene characteristics of texts becomes the point of entry.

1.2 Introducing Narrative into the Metaverse for Scene Animation

Joseph Frank proposed the theory of "spatial form", which belongs to the category of narratology. It is observed that in certain literary texts, narratives primarily unfold around a relatively singular and fixed spatial context [2]. This spatial situation often imitates or reproduces historical or real-life settings, aligning with people's logic of life, thus constituting a substantial and tangible "textual reality space". The selected text for this study, "A Record of Buddhist Monasteries in *Luoyang*", is precisely based on the real spatial foundation of *Luoyang* in *the Northern Wei* Dynasty [19]. Similarly, in modern short stories such as The Furnished Room by O. Henry, narratives unfold around a perceivable real space [5]. Furthermore, the "spatial form" defined by scholars like Joseph is an emotional space generated through the joint interaction of writers and readers, a perceptual space formed through the reader's recreation. Due to the limited description of the real space in the text, the scene layout generated by the text itself necessarily involves AIGC for computer vision supplementation and fitting; this constitutes a virtual space superimposed on the concrete perceptible "real space" of the textual story.

Joseph Frank argues that modernist literature is "spatial", whether in novels, poetries, or others, and the "simultaneity" of space replaces the "sequence" or series of time [18]. It seems that what the novel presents is no longer narration, but a large number of fragmentary details. These detailed presentations manifest a spatial form [17], thereby exhibiting characteristics of temporal suspension and juxtaposed structure based on spatialized scenes.

In this sense, the differences between traditional animation-picture narrative and the spatial narrative that this study aims to investigate are obvious: traditional animation-picture books always tell a story, revealing one aspect and possibility of the narrative object, while animation-picture books based on spatial models tell many stories, revealing multiple aspects of the narrative object and endless possibilities. This model is particularly suitable for new media production and has strong interactivity between audience-readers.

1.3 Overview of Research

The subject of this study is to construct a visual ancient urban space based on historical imagination, aiming to propose a lightweight pathway for generating axonometric vector scene animations from text within the context of the metaverse. The technical methods involved in this research encompass axonometric image representation, computer vision, text semantic analysis, computational geometric modeling, etc. The research methods employed include literature review, interdisciplinary research, survey research, computational research, and so on.

This research is based on the introduction of axonometric representation as a lightweight approach to reconstruct urban spaces and events. It begins by analyzing historical texts to extract spatial data and personal event information related to the city of *Luoyang* in *the Northern Wei* Dynasty. The former focuses on examining the composition, scale, and layout of individual buildings, the distribution of the urban road network, and the relationship between courtyards and streets, estimating and inferring the basic spatial organization of the ancient city. The latter focuses on integrating people and objects into a multi-element framework based on dynastic rituals and historical sources, placing them within the architectural space by considering their relative positions, spatial overlaps, and movements. These two approaches jointly depict the daily life of *Luoyang* in *the Northern Wei* Dynasty, producing a rich collection of animated films portraying the ancient city (see Fig. 16.1).

This study contributes to visually presenting the spatial relationships of ancient urban descriptions from historical texts, promoting historical research. It also aids in the low-cost generation of urban spatial narrative animations, enhancing cultural dissemination. Furthermore, it facilitates the provision of lightweight digital spatial scenes, enriching the development of the metaverse.

2 A Lightweight Metaverse Form: Spatialization of Axonometric Vector Images

Vector graphics use geometric primitives such as points, lines, and simplified shapes to represent images. Their quality remains undiminished after scaling, as they are not resolution-dependent, and the files are easily editable with a smaller file size. According to Gestalt psychology, human perception prefers simplification, which means that the less information required to describe the organization of an object, the more likely it is to be perceived [8]. Simplified vector graphics are suitable for focusing or magnifying the story's setting in metaverse scenes according to narrative demands and can be arbitrarily scaled. The vector format employed in this study, inspired by axonometric drawings, is a "measurable" two-dimensional projection displaying three-dimensional space. It simultaneously reflects the three coordinate planes of an object on a projection plane, closely resembling human visual habits and providing a sense of depth (see Fig. 16.2). This method finds extensive applications in 3D modeling, artistic creation, and computer graphics interfaces.

The article begins with ancient urban texts that have restoration and experiential requirements, transforming the architectural blocks analyzed from the text into combinations of vector lines with computable lengths and locatable coordinates. The lines

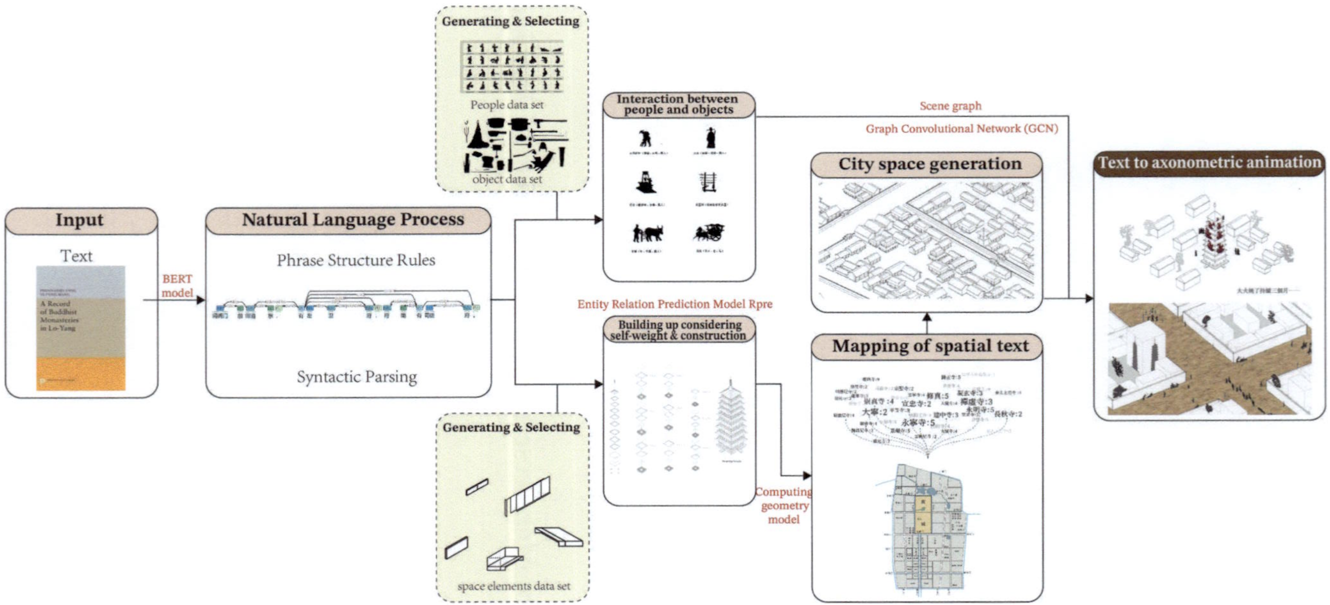

Fig. 16.1 Pathway of text to animated axonometric space

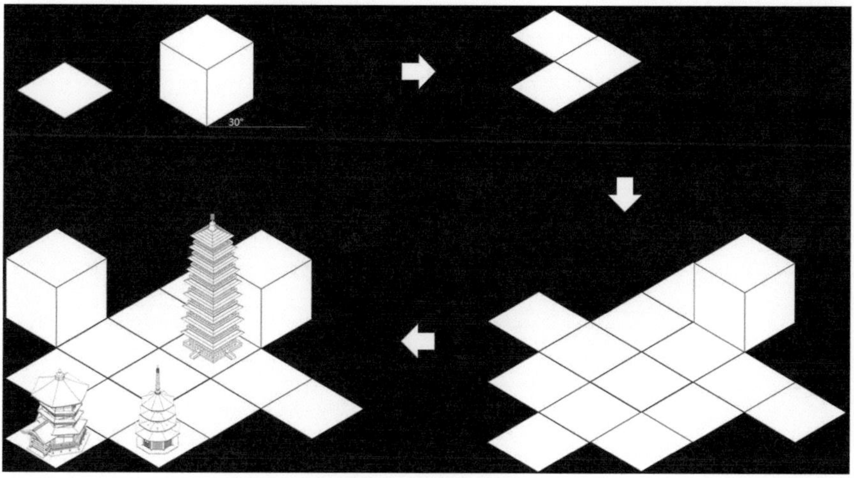

Fig. 16.2 Spatialization of axonometric vector images

adopt a simplified drawing style, emphasizing their different spatial logic relationships, such as progression, overlap, symmetry, etc. Gestalt psychology suggests that closed and meaningful forms are more likely to form shapes [8]. Each textual element is broken down into axonometric representations, expressed in closed vector line drafts, resulting in individual components such as "roofs, beams, columns, walls, bases, steps, ramps", and other architectural elements (see Fig. 16.3). With a predefined baseline, the material's weight and the physical reality effect of solid construction can be presented through height-based stacking.

In painting, when several objects meet locally on a projection plane, they are perceived as having a distinct separation relationship of foreground and background depth. However, if the overlapping parts are attributed to the two objects, the phenomenon of "Transparence"emerges, that commonly seen in modernism, —introducing a new spatial dimension beyond the physical order of reality [15]. Compared to graphics with transparency, axonometric drawings allow for the intuitive understanding of the relative positions between buildings through occlusion relationships (from the size distribution of neighborhood positions to the arrangement of buildings within each block), delineating the cognitive five elements of city image—roads, districtions, landmarks, boundaries, and nodes [11]. This facilitates the restoration of the virtual metaverse space of ancient cities.

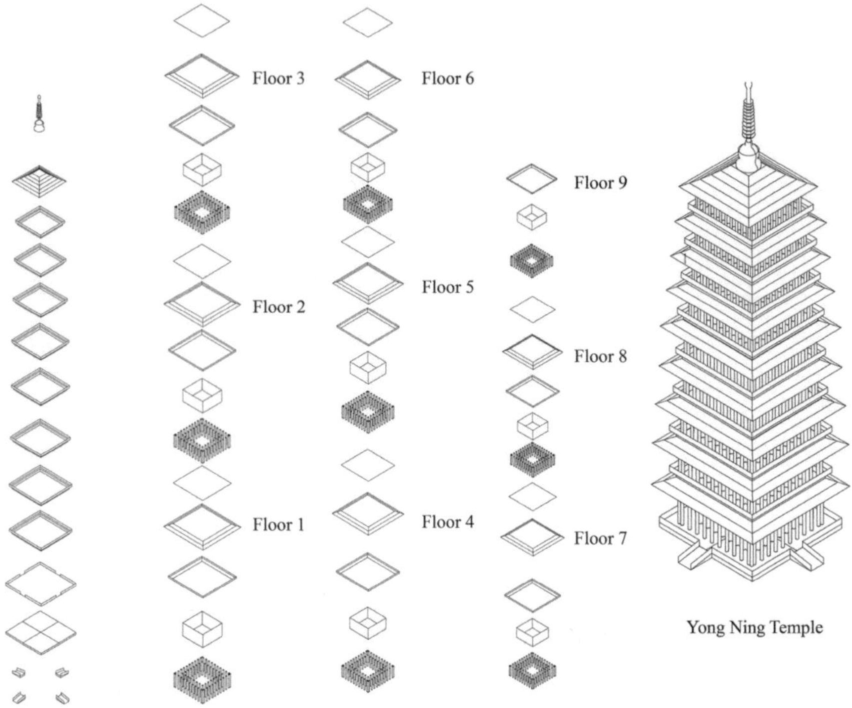

Fig. 16.3 Model of Yongning temple

3 Deconstruction of Text and Correspondence with All Spatial Elements

Gestalt psychology posits that human perception is inherently organized as a coherent whole entity, which means that there are more significances in the whole and the whole entity is not simply a sum of its parts. Humans naturally distinguish between Figure and ground [8]. Therefore, the decomposition and consideration of graphic elements are particularly important, as two-dimensional images are composed of points, lines, shapes, and their combinations. Characters and objects are also essential components of the urban metaverse space. Based on textual records and ancient images, the actions and gestures of people from different dynasties are exhaustively classified (such as hand gestures, sitting postures, bowing, etc.). Objects include key props relevant to the plot and the characteristic images of artifacts from the corresponding dynasties, classified based on categories such as clothing, food, shelter, and transportation. Technically, this study utilizes HanLP model [4] to achieve employing part-of-speech tagging and syntactic analysis tools to parse the text structure, construct semantic trees, and extract the entities and their interaction relationships implied in the text (see Fig. 16.4).

For example, through Abstract Meaning Representation (AMR) analysis of the phrase "*Guo Zi Nan You Zong Zheng Si Si Nan You Tai Miao Miao Nan You Hu Jun Fu Fu Nan You Yi Guan Li*" (see Fig. 16.5), we observe that "*Guo Zi*", "*Zong Zheng*

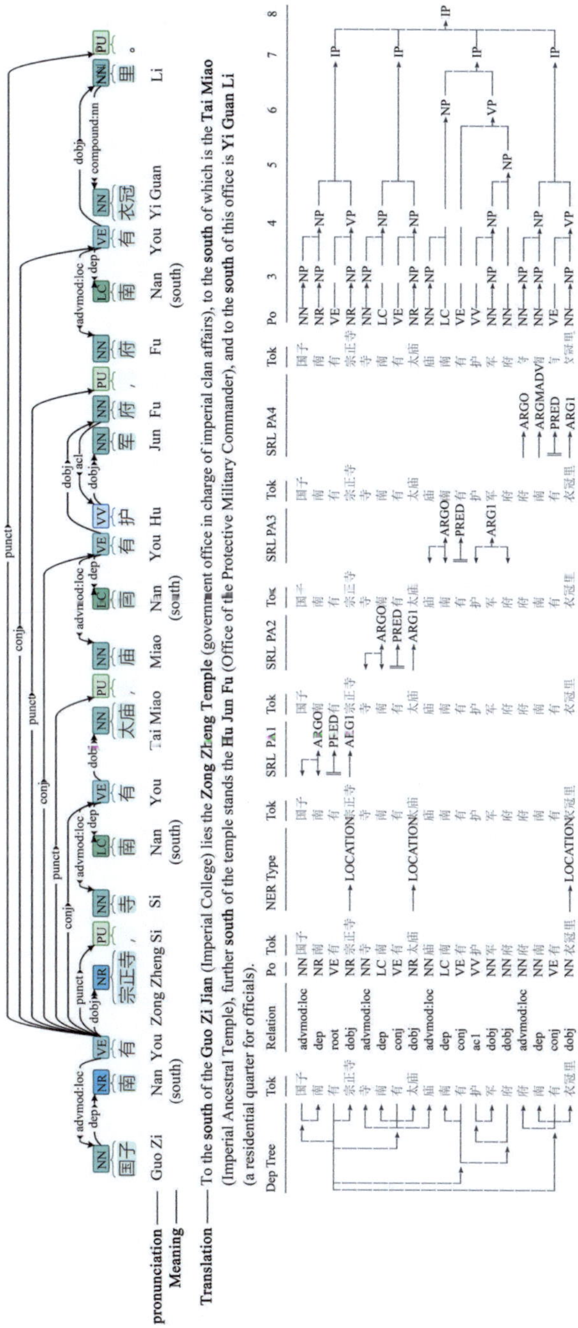

Fig. 16.4 Constituency parsing by HanLP_restful

Si", and "*Tai Miao*" are correctly identified as nouns (NN), and "*Zong Zheng Si*" forms a "domain" relationship with the subsequent "*Si*". However, "*Hu Jun Fu*" is split into "*Hu/Jun/Fu*" due to its verb-object structure, and "*Li*" in "*Yi Guan Li*" is sometimes identified as a spatial relationship and sometimes as a proper noun, depending on the model. Positional terms like "*Nan*(south)" are consistently recognized as location indicators. This series of relative relationships between buildings reveals that "*Zong Zheng Si*" is located to the north of "*Yi Guan Li*", "*Hu Jun Fu*", and "Tai Miao", with each building positioned successively northward and south of "Guo Zi". Based on this, the study expands the historical texts related to *the Northern Wei* Dynasty, supplementing proper nouns and creating a specialized dictionary to improve the accuracy of urban spatial text recognition.

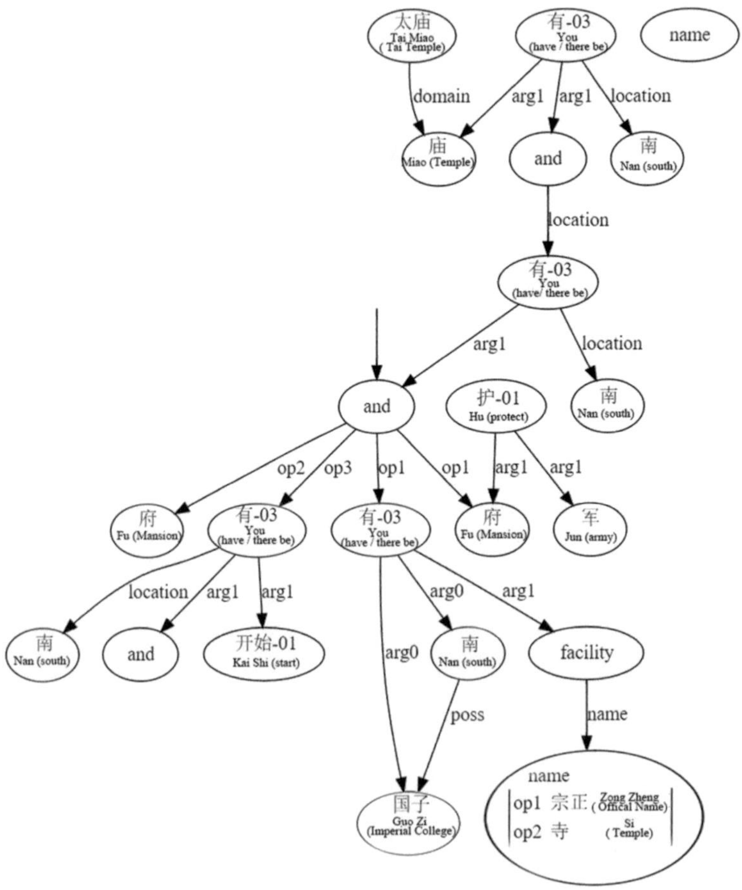

Fig. 16.5 Abstract meaning representation by HanLP 2.0

As described above for "*Zong Zheng Si (Temple)*", "A Record of Buddhist Monasteries in *Luoyang*" mentions more than 70 temples, including "*Yongning Temple, Jianzhong Temple, Changqiu Temple,* et al." This study cross-references multiple texts and uses semantic analysis at the urban scale to verify the relative positions of 48 temples, mapping them to form a basic road network, water system, and temple distribution. With the increasing detail of archaeological findings and site locations for *Luoyang* in *the Northern Wei* Dynasty, the study further refines the reconstructed urban space using modern Geographic Information Systems (GIS) satellite data.

Moreover, semantic analysis is applied at the block scale to refine the layout of courtyard buildings. Descriptions such as "*Fu Tu You Si Mian, Mian You San Hu Liu Chuang* (the pagoda has four sides, each with three doors and six windows)" and "*Shi Er Men Er Shi Si Shan* (twelve gates and twenty-four doors)" provide module information corresponding to the construction hierarchy of temples, enabling estimations of building sizes and typologies. The types are then matched to archetypes in a typological database, such as linear, L-shaped, C-shaped, or rectangular forms. Descriptions like "*Fo Dian Yi Suo, Xing Ru Tai Ji Dian* (a Buddha hall resembling the Taiji Hall)" and "*Seng Fang Lou Guan Yi Qian Yu Jian* (more than a thousand monk rooms and towers)" depict the type and structure of individual buildings. Additionally, the layout of courtyards and streets is outlined, for example, with "*Si Mian Ge Kai Yi Men. Nan Men Lou San Chong, Tong San Dao, Qu Di Er Shi Zhang, Xing Zhi Shi Jin Duan Men* (each side has a gate, the southern gate has three stories, with three passages, 20 zhang above the ground, resembling the present-day Duan Gate)". Through the analysis of the above historical texts, deductions can be made to estimate the basic spatial landscape of the ancient city of *Luoyang* in *the Northern Wei* Dynasty (see Fig. 16.6).

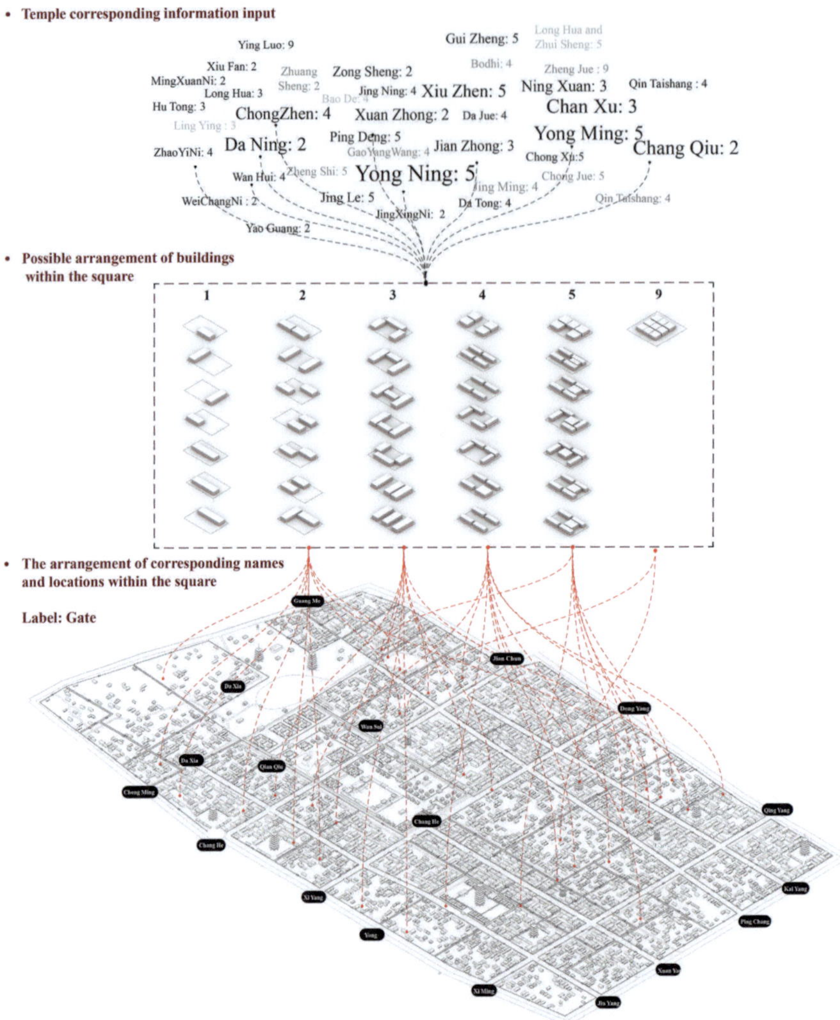

Fig. 16.6 Phrase-driven generative framework

4 Coupling of Image Elements with Scene Space

Gestalt psychology posits that perception tends towards controlling multiple stimuli to form organic entities—this tendency for stimuli to be perceived as a unified whole is referred to as the principle of grouping [8]. Constructivist theory suggests that people from different cultural backgrounds have significant differences in their perceptual biases toward depth information in perspective views. Some Africans perceive that the spears point towards the elephant, which appears closer in two dimensions, while those with modern education are aware that, under the effect of perspective, the elephant is located on a distant mountain, with the spear pointing towards the antelope [10]. However, when the same elements are redrawn using axonometric projection, perception can discern that each element in the image is constrained by a set of axonometric reference lines at 120°angles [16], allowing for a clear differentiation of the size, distance, orientation, and other interactive relationships among the elements in the scene (see Fig. 16.7).

Fig. 16.7 The discrepancy in spatial representation between perspective views and axonometric projections. (Figure source: 4-a, reference 13; 4-b, author-drawn)

This study designs a semantic feature extraction model for the ancient text "A Record of Buddhist Monasteries in *Luoyang*" and selects the datasets for text-to-image generation. Characters and artifacts from *Bei Wei* Dynasty exhibit distinctive characteristics of the era. Accordingly, this study similarly depicts their clothing, hairstyles, and everyday utensils according to the logic of the real world, including wide outer garments, the upper garments and trousers, belts, high-toothed wooden clogs, '*Guo Yi*' (a ceremonial dress in the *Wei* and *Jin* dynasties), agricultural tools, carts, walking sticks, inkstones, bamboo slips, and dyeing utensils[1] (see Fig. 16.8). All of these form a system of pictorial elements that can flexibly complement those above architectural spatial context.

Fig. 16.8 Typical elements of daily ancient life and combination of people actions

Axonometric drawings, devoid of perspective distortion, represent a relatively high-dimensional modality of scene space information based on two-dimensional low-dimensional images. When the viewpoint shifts, the lines of each block remain undistorted, allowing for the creation of a shifting lens effect through the movement of the frame, as well as the ability to move elements such as graphical blocks to advance the development of animated plots. Axonometric urban spatial scenes intuitively embody the "spatial form" of texts by presenting juxtaposed structures, echoing the "scattered point perspective" technique seen in classical Chinese paintings such as the "*Qing Ming Shang He Tu*" and panoramic bird's-eye views. In addition to simple undistorted camera translations, spatial meanings can be generated through techniques such as "merging", "interlacing", and "crossing" of characters and objects in planar combinations (see Fig. 16.9), creating spatial narrative effects such as placing items at specific elevations and people ascending stairs.

Fig. 16.9 Movable elements such as graphical blocks

Generating complex scenes with multiple interacting objects in images is a high-level computer graphics task. It often involves using semantic layouts as one of the steps in the image generation process, predicting the bounding box positions and categories of each entity in the target image beforehand, and utilizing techniques like "graph convolution" to generate scene layouts [6]. This study proposes a phrase-driven generative framework that encodes all words describing the same object in the input text into phrase features. It establishes correspondences between phrases and objects and models relative positional relationships between objects based on phrases [7]. With the aid of scene layout graphs, it is possible to intuitively model various entity objects and their relationships, thereby helping to bridge the semantic gap between text modality and image modality.

If the static axonometric scene graph places spatial elements refer to the spatial formal characteristic of "simultaneity", then when the elements of the axonometric scene are moved and coordinated with camera translation and scaling, similar to storyboard, a sequential depth sequence in time can be generated, thereby obtaining a multi-faceted panoramic spatial narrative [3]. Taking human figures as an example, while ensuring minimal loss of semantic transmission, dynamic actions of individuals are efficiently displayed using as few frames as possible. By accumulating a large library of elements (buildings, characters, objects, etc.) in the scene, even with occlusion relationships, all elements remain independently complete, forming definite relationships. For example, "opening a door" may lead to "the appearance of light (inside the room)", facilitating the future invocation of these elements in simple programming to achieve a one-change-all effect (see Fig. 16.10). This allows the ancient urban space to conform to basic architectural logic and initial image intelligence, enabling the narration of architectural and spatial stories. Due to the avoidance of perspective viewpoints and light and texture rendering in three-dimensional axonometry, computational complexity and file size can be greatly reduced, further enhancing the production efficiency of mapping text to lightweight metaverse spaces (see Fig. 16.11).

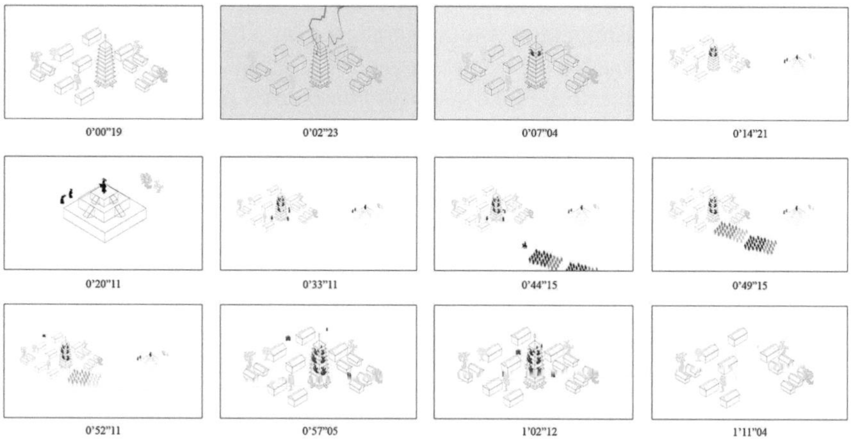

Fig. 16.10 Animation generation about Yongning temple on fire in Luoyang

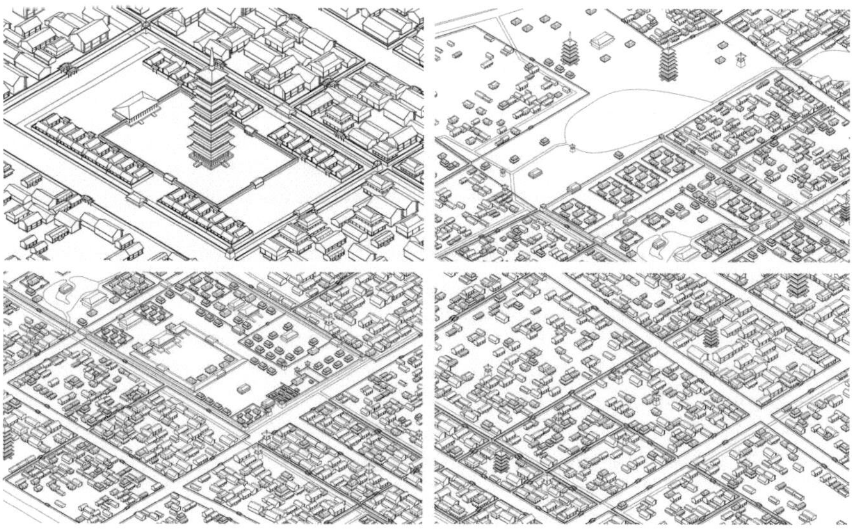

Fig. 16.11 Generation of ancient urban space

5 Conclusion

This study introduces the concept of "spatial form" from narratology and draws upon visual perception theories such as Gestalt psychology and structural psychology. The research focuses on vectorized axonometric representation and takes ancient urban space as a case study to explore a pathway for generating spatial animated metaverse. Technically, the study utilizes the HanLP model and employs part-of-speech and syntactic analysis tools to parse the text structure, constructing a semantic tree to extract implicit entity objects and their interaction relationships from the text. Additionally, the research selects and expands the dataset for text-to-image generation to build a suitable text generation image dataset. Through a phrase-driven generative framework, relative positional relationships between objects are modeled based on phrases, thereby obtaining scene layout diagrams with embedded vectors to bridge the semantic gap between text and image modalities.

By constructing and updating buildings, the rise and fall of cities can be depicted [9], and by depicting changes in characters and objects, humanistic scenes in the city can be deduced. Overlaying movements and scaling of scenes, image-animation based on spatial models can juxtapose multiple narratives, revealing multiple facets and infinite possibilities of the narrative objects, ultimately constituting a lightweight urban metaverse space.

Acknowledgements. The authors gratefully acknowledge the financial support provided by the China Postdoctoral Science Foundation (2023M741955).

References

1. Chen W (2012) On the accurate restoration of historical costumes in ancient costume animated films. Decorations 6:64–65
2. Frank J (1945) Spatial form in modern literature: an essay in two parts. Sewanee Rev 53(2):221–240
3. Furniss (2008) The animation bible: a guide to everything, from flipbooks to flash. Laurence King
4. He H, Choi JD (2021) The stem cell hypothesis: Dilemma behind multi-task learning with transformer encoders. arXiv preprint arXiv:2109.06939
5. Henry O (2020) The furnished room. Lindhardt og Ringhof
6. Hong S, Yang D, Choi J, Lee H (2018) Inferring semantic layout for hierarchical text-to-image synthesis. In: Proceedings of the IEEE conference on computer vision and pattern recognition, pp 7986–7994
7. Johnson J, Gupta A, Fei-Fei L (2018) Image generation from scene graphs. In: Proceedings of the IEEE conference on computer vision and pattern recognition, pp. 1219–1228
8. Koffka K (1935) Principles of gestalt psychology, 1st edn. Harcourt, Brace, New York
9. Kostof S (1991) The city shaped, pp 9–39. Little, Brown and Company
10. Krech D, Crutchfield RS, Livson N (1974) Elements of psychology. Knopf, Alfred A
11. Lynch K (1964) The image of the city. MIT Press
12. Ramesh A, Pavlov M, Goh G, Gray S, Voss C, Radford A,... Sutskever I (2021) Zero-shot text-to-image generation. In: International conference on machine learning, pp 8821–8831 PMLR

13. Reed S, Akata Z, Yan X, Logeswaran L, Schiele B, Lee H (2016) Generative adversarial text to image synthesis. In: International conference on machine learning, pp 1060–1069. PMLR
14. Rombach R, Blattmann A, Lorenz D, Esser P, Ommer B (2022) High-resolution image synthesis with latent diffusion models. In: Proceedings of the IEEE/CVF conference on computer vision and pattern recognition, pp 10684–10695
15. Rowe C, Slutzky R (1963) Transparency: literal and phenomenal. Perspecta, 45–54
16. Thomas H (2018) Drawing architecture. Phaidon
17. Wu X (2017) From Kafka to Kundera: Novels and novelists of the 20th century. Sanlian Bookstore, Life·Reading·New Knowledge
18. Wu Y (2008) Reproduction of spatial theory and literature. Gansu People's Publishing House
19. Yang X, Zhou Z (2022) Luoyang Jialan Records. Zhonghua Book Company

MIX
Papier aus verantwortungsvollen Quellen
Paper from responsible sources
FSC® C105338

If you have any concerns about our products,
you can contact us on
ProductSafety@springernature.com

In case Publisher is established outside the EU,
the EU authorized representative is:
Springer Nature Customer Service Center GmbH
Europaplatz 3, 69115 Heidelberg, Germany

Printed by Libri Plureos GmbH
in Hamburg, Germany